ENERGY SCIENCE, ENGINEERING AND TECHNOLOGY

SOLAR WIND

EMISSION, TECHNOLOGIES AND IMPACTS

ENERGY SCIENCE, ENGINEERING AND TECHNOLOGY

Additional books in this series can be found on Nova's website under the Series tab.

Additional E-books in this series can be found on Nova's website under the E-book tab.

PHYSICS RESEARCH AND TECHNOLOGY

Additional books in this series can be found on Nova's website under the Series tab.

Additional E-books in this series can be found on Nova's website under the E-book tab.

ENERGY SCIENCE, ENGINEERING AND TECHNOLOGY

SOLAR WIND

EMISSION, TECHNOLOGIES AND IMPACTS

CARLOS DANIEL ESCAROPA BORREGA
AND
ANGELA FERNANDA BEIRÓS CRUZ
EDITORS

Nova Science Publishers, Inc.
New York

For permission to use material from this book please contact us:
Telephone 631-231-7269; Fax 631-231-8175
Web Site: http://www.novapublishers.com

NOTICE TO THE READER

Additional color graphics may be available in the e-book version of this book.

Library of Congress Cataloging-in-Publication Data

Solar wind : emission, technologies, and impacts / editors, Carlos Daniel Escaropa Borrega and Angela Fernanda Beirss Cruz.
p. cm.
Includes bibliographical references and index.
ISBN 978-1-62081-979-1 (soft cover)
1. Solar wind. I. Escaropa Borrega, Carlos Daniel. II. Beirss Cruz, Angela Fernanda.
QB529.S6255 2011
523.5'8--dc23
2012017832

Published by Nova Science Publishers, Inc. † New York

CONTENTS

Preface		**vii**
Chapter 1	Possible Impact of the Astronomical Aspects on the Violent Cyclonic Motions in the Earth's Atmosphere *Joao Fernando Pereira Gomes, Saumitra Mukherjee, Milan M. Radovanović, Boško Milovanović, Luka Č. Popović and Andjelka Kovačević*	**1**
Chapter 2	Planetary Beat, Solar Wind and Terrestrial Climate *Nils-Axel Mörner*	**47**
Chapter 3	Was the Idea of Solar Wind Wafting Around Even Before the Parker Formulation? *R. P. Kane*	**67**
Chapter 4	Solar Wind Influence on Atmospheric Processes in Winter Antarctica *O. A. Troshichev, V. Ya. Vovk and L. V. Egorova*	**73**
Chapter 5	Experimental and Modeling Evidences of the Solar Wind Energy Influence on the Earth Atmosphere *L. N. Makarova and A. V. Shirochkov*	**107**
Index		**135**

PREFACE

The continuous stream of particles flowing outward from the Sun is called the solar wind. It mostly consists of electrons and protons with energies usually between 1.5 and 10 keV. The stream of particles varies in temperature and speed over time. These particles can escape the Sun's gravity because of their high kinetic energy and the high temperature of the corona. In this book, the authors present current research in the study of the emission, technologies and impact of solar wind. Topics include the possible impact of the astronomical aspects of the violent cyclonic motions in the Earth's atmosphere; planetary beat, solar wind and terrestrial climate; historical background on Eugene Parker and his conception and formulation of solar wind; solar wind influences on atmospheric processes in winter Antarctica; and experimental and modeling evidences of solar wind energy on the Earth's atmosphere.

Chapter 1 - The process of forming cyclonic motions in the atmosphere (cyclones, depressions, storms, hurricanes, etc.) has been the subject of numerous studies. However, the mechanisms of origin of this phenomenon are not well known. Not only that there are doubts about the way they are formed, but also about their development in time and space. In general, prognostic methods, with a greater or lesser success, are applied only at the moment when there is a disturbance in the atmosphere. In this paper the authors discussed the possible effects of astronomical aspects on the occurrence of cyclones in the earth's atmosphere. Firstly, an overview of potential impacts in regard to the movement of the earth is presented, and then the importance of the solar wind (SW) in the particular and global earth's atmospheric motions discussed. Recently, a number of scientific studies have been focused on the connection between the parameters of the SW and atmospheric disturbances. In contrast to common interpretations, both in the media and the wider scientific public, that direct or indirect anthropogenic activity is the cause of these processes, the potential links are explained primarily in the context of the energy arriving from the Sun. The focus of presentation in the following pages is directed primarily to the research results of a recent date. The authors hope that the discussed literature may be of great help to all those who want to study different approaches of cyclogenesis.

Chapter 2 - The Sun's emission of luminosity and Solar Wind is constantly changing. This solar variability is driven by a planetary beat because the Solar-Planetary system behaves like a multi-body system constantly adjusting their motions around a common centre of gravity. The solar influence on terrestrial climate and environments seems primarily to go via the interaction of the Solar Wind with the Earth's magnetosphere. This generates changes in shielding capacity, geomagnetic activity, atmospheric pressure, external gravity and Earth's rate of rotation. The correlation between changes in solar activity and Earth's rate of rotation is indicative of a forcing function via changes in Solar Wind emission (not luminosity). Changes in rotation (LOD) affects not only the atmospheric circulation but also the ocean circulation. Changes in ocean circulation lead to the redistribution of oceanic water masses (recorded by sea level changes) and water-stored heat (recorded by paleoclimate). Changes in ocean circulation have strong effects on terrestrial climate. During Grand Solar Minima (the Spörer, Maunder and Dalton Minima) the Earth's experienced a speeding-up in the rate of rotation as evidenced by significant reorganisations of the currents in the North Atlantic. This lead to the establishment of periods of cold climatic conditions, known as Little Ice Ages. The next Grand Solar Minimum is due at about 2030-2040. We must assume that this will be a period of resumed cold climatic conditions, may be even Little Ice Age conditions.

Chapter 3 - The idea of solar wind was conceived and formulated by Eugune Parker in a rigorous way in mid 1950s. However, it seems that the idea was floating around in scientific circles even before in some form or other.

Chapter 4 - The paper presents a summary of the experimental results demonstrating the strong influence of the interplanetary electric field on atmospheric processes in the central Antarctica, where the large-scale system of vertical circulation is formed during the winter seasons. The influence is realized through acceleration of the air masses, descending into the lower atmosphere from the troposphere, and formation of cloudiness above the Antarctic Ridge, where the descending air masses enter the surface layer. The cloudiness formation results in the sudden warmings in the surface atmosphere, since the cloud layer efficiently backscatters the long wavelength radiation from the ice sheet, but does not affect the adiabatic warming process of the descending tropospheric air masses. The acceleration is followed by a sharp increase of the atmospheric pressure in the near-pole region, which gives rise to the katabatic wind strengthening above the entire Antarctica. As a result, the circumpolar vortex about the periphery of the Antarctic continent correspondingly decays and the cold air masses flow out to the Southern ocean. The latter phenomena evidently destroys the regular relationships between the sea level pressure fluctuations in the Southeast Pacific high and the North Australian-Indonesian low. It seems that the El-Niño beginnings are related exactly to the anomalous atmospheric processes in the winter Antarctica.

Chapter 5 - So far the solar wind energy contribution to energetic balance of the Earth atmosphere was ignored in any atmospheric and climatic research. However the solar wind is a permanent source of a significant amount of the electromagnetic energy emitted by the Sun which is constantly supplied to the near-Earth space. Traditionally this energy was attributed entirely to sustain a definite level of geomagnetic activity expressed as intensity of the geomagnetic substorms and storms. The authors of this paper found in 1997 after analysis of the data of the Russian rocket sounding in the Arctic that enhancement of the solar wind dynamic pressure do influence thermal regime of the polar middle atmosphere. Similar analysis of the atmospheric balloon sounding data obtained at different stations in both the Arctic and the Antarctica shows that the stratospheric temperature closely correlated with the solar wind electromagnetic energy. After establishing these statistically confident relations it was necessary to find a plausibly reasonable physical mechanism which could explain reality of the found coupling. A concept of the global electric circuit as a physical mechanism for explanation of a direct coupling between the solar wind and the middle atmosphere was suggested. We proposed a new, modified version of the global electric circuit with two Electro-Motive Force (EMF) generators: internal EMF generator driven by the thunderstorm activity of the Earth (a common feature of previous circuit configurations) and an external EMF generator driven by the solar wind energy. The passive elements of this circuit are the ionospheric E-layer (external element of previous version of the circuit), stratospheric conducting layer of heavy ions (h=20-25 km) and conducting layer of the Earth surface. In this configuration a previous scheme of the global electric circuit is a part of the proposed version of it. Numerical evaluation of the electromagnetic energy of the solar wind is a very difficult task. It can be done only approximately. Structure of the Earth magnetosphere is changing constantly upon influence of the solar wind as well as a position of a boundary of the magnetosphere (magnetopause). The problem could be solved if we present boundary of the Earth magnetosphere and the ground surface as a giant capacitor with external and internal plates correspondingly. The external plate of this capacitor (magnetopause) could be moved toward the Earth under the solar wind pressure. The energy of the solar wind roughly can be calculated by estimation of energy which is required to move the magnetopause for a definite distance. The magnetopause is located at ~ 12 Re (were Re is the Earth radius) under a quite condition of the solar wind. During strong disturbances of the solar wind the magnetopause could approach the Earth at distance of approximately ~ 6 Re. Such estimation shows that energy required for movement of magnetopause at a distance of 6 Re is equal to ~ 5 10 15J. Preliminary numerical estimations showed that under typical conditions such amount of the Joule heating dissipated in stratosphere is comparable with a rate of heating of ozone layer by the solar UV radiation. Furthermore, such amount of energy is sufficient for enhancement of cyclonic activity in the Earth atmosphere. As the next step of exploration a numerical calculation scheme was elaborated, which took into account the abovementioned processes. This numerical scheme was successfully used in one of the global dynamical photo-chemical models of the atmospheric circulations. The results of these model simulations confirmed all previously made preliminary estimations concerning influence of the solar wind energy on the atmospheric processes. There are the definite plans to improve the effectiveness of the proposed physical mechanism describing interaction of the solar wind with the Earth atmosphere. Evaluation of the effects of different degree of the Earth electric conductivity must be taken into account in the next explorations on the subject.

In: Solar Wind ISBN 978-1-62081-979-1
Editors: C. D. E. Borrega, A. F. B. Cruz

Chapter 1

POSSIBLE IMPACT OF THE ASTRONOMICAL ASPECTS ON THE VIOLENT CYCLONIC MOTIONS IN THE EARTH'S ATMOSPHERE

Joao Fernando Pereira Gomes[1,*], Saumitra Mukherjee[2], Milan M. Radovanović[3], Boško Milovanović[3], Luka Č. Popović[4] and Andjelka Kovačević[5]

[1]Chemical Engineering Department, IST — Instituto Superior Técnico, Torre Sul, Lisboa, Portugal and Chemical Engineering Department, ISEL — Instituto Superior de Engenharia de Lisboa, R. Conselheiro Emídio Navarro, Lisboa, Portugal

[2]School of Environmental Sciences Jawaharlal Nehru University, New Delhi, India

[3]Geographical Institute "Jovan Cvijic", Serbian Academy of Sciences and Arts — SANU, Belgrade, Serbia

[4]Astronomical Opservatory Belgrade, Belgrade, Serbia

[5]Department for Astronomy, Faculty for Mathematics, Belgrade, Serbia

ABSTRACT

The process of forming cyclonic motions in the atmosphere (cyclones, depressions, storms, hurricanes, etc.) has been the subject of

* E-mail: jgomes@deq.isel.ipl.pt.

numerous studies. However, the mechanisms of origin of this phenomenon are not well known. Not only that there are doubts about the way they are formed, but also about their development in time and space. In general, prognostic methods, with a greater or lesser success, are applied only at the moment when there is a disturbance in the atmosphere. In this paper we discussed the possible effects of astronomical aspects on the occurrence of cyclones in the earth's atmosphere. Firstly, an overview of potential impacts in regard to the movement of the earth is presented, and then the importance of the solar wind (SW) in the particular and global earth's atmospheric motions discussed. Recently, a number of scientific studies have been focused on the connection between the parameters of the SW and atmospheric disturbances. In contrast to common interpretations, both in the media and the wider scientific public, that direct or indirect anthropogenic activity is the cause of these processes, the potential links are explained primarily in the context of the energy arriving from the Sun. The focus of presentation in the following pages is directed primarily to the research results of a recent date. We hope that the discussed literature may be of great help to all those who want to study different approaches of cyclogenesis.

Keywords: solar wind, charged particles, cyclonic motions

1. INTRODUCTION

The violent cyclonic motions in the Earth's atmosphere, such as hurricanes, cannot be predicted and it is yet not clear what is the mechanism of their creation and development. Hurricanes are one of the most violent cyclonic motions in the atmosphere. They represent extremely strong and destructive disasters that endanger the coastal areas, particularly in the tropical belt. Their relatively sudden appearance endangers both the sea and air transportation, and besides considerable material losses, they also take away many victims. For a long time scientists have searched for the causes of their occurrence, as well as possible explanation of the mechanism on the basis of which they are functioning. Love (2006) emphasizes that the few available evidence points out to an expectation of little or no change in the global frequency of hurricane appearance. Regional and local frequencies could change substantially in either direction, because of the dependence of cyclone genesis and track on other phenomena (e.g. El Nino Southern Oscillation - ENSO) that are not yet predictable. Observations from Markowski and

Richardson (2009) do not differ much from the previously quoted authors. According to them, there are a number of aspects of super cell thunderstorms and tornado genesis that remain poorly understood. The temporal and spatial variations in storm activity are quite different for weaker tropical cyclones (TC-tropical storm up to category 2 in strength) than for stronger storms (categories 3–5). The stronger storms tend to show stronger inter-basin correlations and stronger relationships to ENSO and the North Atlantic Oscillation (NAO) than do weaker storms. This suggests that the factors that control tropical cyclone formation differ in important ways from those that ultimately determine storm intensity (Frank and Young, 2007). Barrett and Leslie (2009) also found similar results, concluding that upper-tropospheric divergence is the physical link between TC activity and the Madden–Julian oscillation (MJO). The eastward-propagating Kelvin wave sequentially modulates large-scale upper-tropospheric conditions, which impact TC genesis and intensification.

The opposite ideas on the link between anthropogenic activity and cyclones can very often be found in scientific publications. While the number and timing of storms of tropical origin are likely to increase, this increase appears to be attributed to a multi-decadal cycle, opposite to a trend in global warming (Vermette, 2007). Understanding the physics of the cyclonic motion involves a complex knowledge of fluid dynamics, thermodynamics and scale of interactions. However, approaches presented in this chapter do not have any connection with the anthropogenic activity. Instead, the potential links have primarily been investigated in the context of energy coming from the Sun and its influence on possible violent cyclonic motion in the Earth's atmosphere.

In particular, Hocke (2009) has emphasized that the solar wind quasi-biennial oscillation (QBO) may influence the stratospheric QBO, the global electric circuit, and cloud cover by modulation of ionospherical electric fields, cosmic ray flux, and particle precipitation. Georgieva *et al.*, (2007) have concluded that the long-term correlation between solar activity and atmospheric circulation changes in consecutive secular solar cycles and also depends on the north–south asymmetry of solar activity: when the northern solar hemisphere is more active, the increasing solar activity in the secular (Gleissberg) cycle leads to decreasing prevalence of zonal forms of circulation. On the other hand, when the southern solar hemisphere is more active the increasing solar activity in secular solar cycles leads to increasing zonality of atmospheric circulation. Moreover, the causative-effective links were explained between solar electromagnetic radiation, solar corpuscular radiation, solar flares, solar coronal mass ejections, magnetic field perturbations, high speed solar wind and types of general circulation.

Some facts about solar influence on Earth's atmosphere, which could be the basis for understanding the way of the origin of cyclones, especially hurricanes, are presented in this chapter. In section 2 we describe the astronomical aspects of the Sun-Earth connection in the frame of the origin of the most violent cyclonic motion such as hurricanes. In section 3 we present some ideas referring, in particular, several violent cyclonic motions, and finally in section. 4 we shortly outline our conclusions and perspectives for future investigations in this field.

2. Astronomical Aspects of Violent Cyclonic Motions

Here we discuss, in more detail, the astronomical and astrophysical aspects of solar radiation as a possible cause of violent cyclonic motion in the Earth's atmosphere. For a certain place on Earth, the amount of solar radiation received depends on several, but two are the most important ones: i) astronomical aspects caused by the revolution and rotation of Earth, and ii) astrophysical aspects connected with the Sun processes, taking into account the complex physics of the Sun-Earth system. We will take into account both aspects, but before that, let us consider some global characteristics of hurricanes which can be connected with astronomical aspects.

2.1. Some Global Characteristics of Hurricanes – Places of Origin and Time of Beginning

One of the most important parameters in initiating a cyclonic motion is the energy supplied from the Sun. Therefore, we will start by considering the obtained solar energy for a certain geographical coordinate on Earth where violent cyclonic motions are starting.

It is well known that violent cyclonic motions start around equatorial regions on the Earth. Figure 1 shows places in the North hemisphere where hurricanes were detected in the period from 1997 to 2010 (data for 404 hurricanes from the Atlantic and Pacific are taken from the database of the Atlantic Oceanographic and Meteorological Laboratory).

It can be seen, from this figure, that the latitude of places where hurricanes are starting is mainly below 45 degrees, and the highest number of hurricanes, in the past 14 years, started between 10 and 30 degrees (around 90). Also, in the West longitude, most hurricanes were starting between -10 and -120 degrees, and in the East part there is a narrower interval from 110 to 180 degrees.

Of course, the places of origin are located mainly in oceans, i.e. above a certain amount of the ocean water mass, but some additional conditions should be reached at the moment of the starting of a hurricane.

The relevant fact is that the radiation coming from the Sun is changing during one year due to the Earth revolution (see section 2.2). Therefore, during a year the input of the energy coming from the Sun for a certain geographical coordinate will be different.

Considering the North hemisphere, figure 2 shows the plot of the monthly distribution of hurricanes in the period of 1997-2010. The data are similar as the ones presented in figure 1.

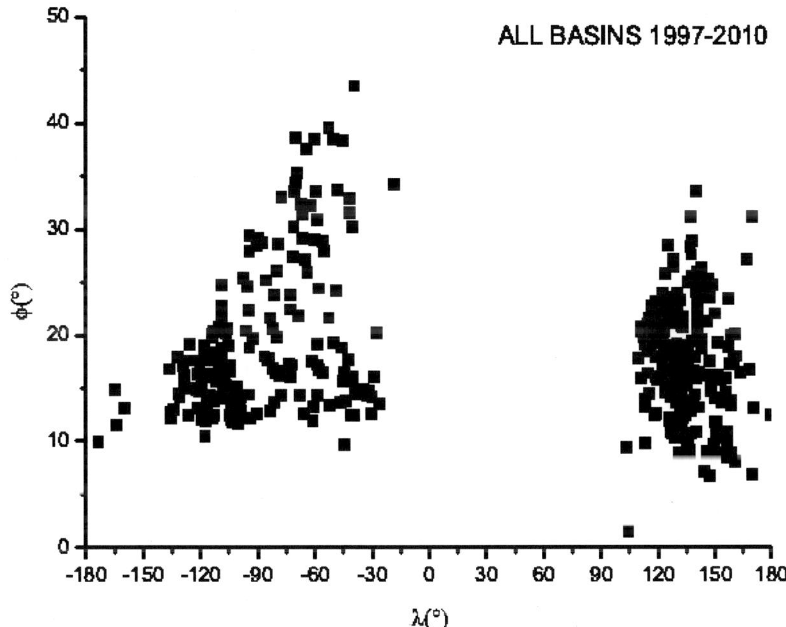

(Source: http://www.aoml.noaa.gov/hrd/data$_$sub/hurdat.html).

Figure 1. Places on Earth where hurricanes were starting (latitude and longitude). The data are from Atlantic, West and East Pacific and are covering the period from 1997 to 2010.

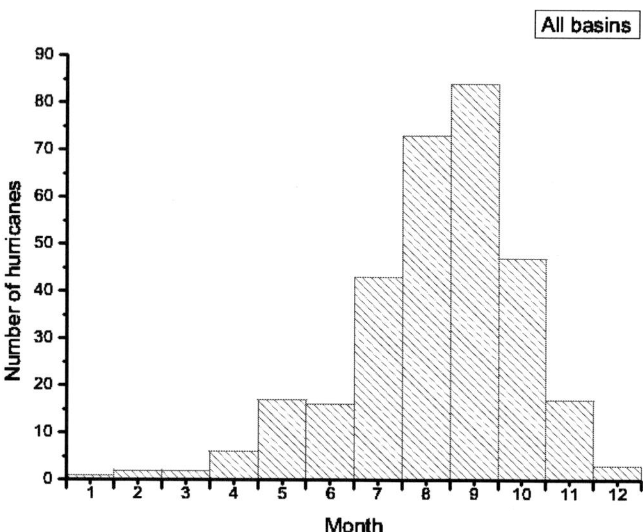

The data are taken from the same source as in figure 1.

Figure 2. Monthly distribution of hurricanes in the period 1997-2010, where a maximum of number of hurricanes is noticeable in August-September.

As it can be seen (and is also well known), most hurricanes appear in summer time, but also there is a significant number of hurricanes which were detected in September and October. In any case, more than half of them appeared between spring and autumn Equinox, which may be also connected with some type of solar radiation.

To discuss the violent cyclonic motions, or hurricanes, it is necessary to take into account the complex Earth's atmosphere and different processes between the atmosphere and hydrosphere. However, we could pose one question: is there any indication that some of the additional parameters caused by the solar radiation may cause violent cyclonic motion in the Earth's atmosphere? To discuss this, let us first recall some relatively well known facts about the solar radiation and its influence on the Earth.

2.2. Daily and Yearly Solar Irradiation of the Earth Surface and Hurricane Appearances

Starting from the point that the Earth's atmosphere is heated by the solar energy, one should consider how much the variation in the daily and yearly obtained energy can influence the turbulence in the Earth's atmosphere.

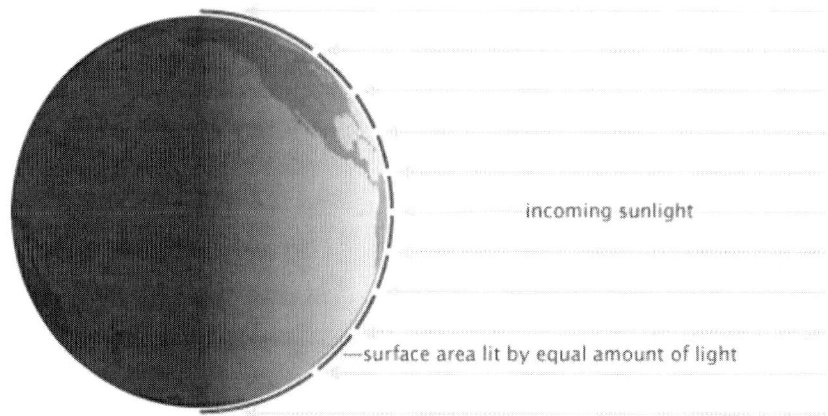

Figure 3. Different energy received from the Sun in time depends on the geographical longitude. The Equatorial belt receives more energy.

First, let us consider that different places on Earth will receive different amounts of Solar energy due to the inclination of the rays coming from the Sun. Solar radiation is unevenly distributed throughout the world because of such variables such as solar altitude, which is associated with latitude and season, and also atmospheric conditions (which are determined by the cloud coverage and degree of pollution).

As can be seen in figure 3, there are some specific geographic areas based on the collection of the direct component of sunlight. One can expect that the equatorial belt (between 0-15° North and South) gets the most sunlight. This is true if one considers it from the top of the Earth's atmosphere. However, the equatorial belt has high atmospheric humidity and cloudiness that tends to increase the proportion of the scattered radiation. The global solar intensity is almost uniform throughout the year, as there are only slight seasonal variations. Sunshine is estimated to reach 2500 h/year (Acra *et al.* 1990). On the other hand, the equatorial belt, between 15-35° North and South, reaches more than 3000 h/year of sunshine and limited cloud coverage, i.e. more than 90% of the incident solar radiation comes as direct radiation. As it can be seen in figure 1, this belt is favourable for hurricanes origin, and, in fact, most of hurricanes have been detected in this belt.

In the longitudes higher than 35° North and South, the scattering of the solar radiation is significantly increased because of the higher latitudes and lower solar altitude. In addition, cloudiness and atmospheric pollution are important factors that tend to reduce sharply the solar radiation intensity.

Moreover, in the region beyond 45° North and South, almost half of the solar radiation is in the form of scattered radiation.

Considering the highest layers of the Earth's atmosphere, it is well known that the amount of received energy per unit time per area on Earth depends on the Sun position in the sky for an observer situated in this place. Due to the rotation of Earth the most energy will be collected when the Sun is in culmination (crossing the local meridian). On the other hand, during the year, it depends on the Earth's tilt, and as it can be seen in figure 4: the most energy is cumulated in North Hemisphere in summer time. But that can be explained with the fact that the energy from the Sun is mostly collected in the period of year which overlaps the period of hurricanes (see figure 2).

Taking into account all facts considered above, one can conclude that the variability in solar irradiation of the Earth's atmosphere and surface due to the Earth revolution and rotation has significant influence on violent cyclonic motion.

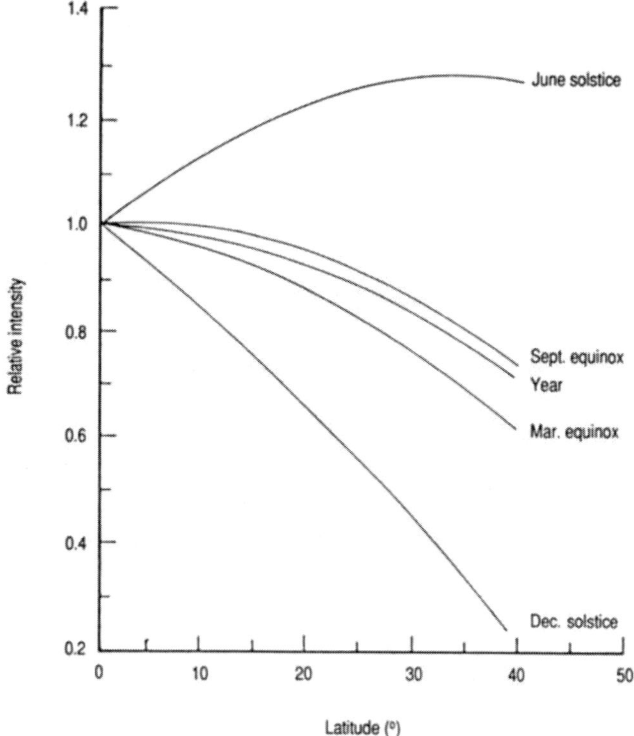

Figure 4. Seasonal and annual variations in the relative solar UV radiation at the wavelength of 340 nm for different latitudes (Johnson *et al.*, 1976).

The frequency of hurricanes in Pacific and Atlantic (see figure 1) is higher during summer time, and they are located within longitudes (mostly between 15 and 35 degrees) where more than 90% of the incident daily solar radiation comes as direct radiation.

2.3. Astrophysical Aspects – Perturbations in Solar Radiations and Solar Wind

The second approach to the possible astronomical aspects of violent cyclone motion in the Earth's atmosphere considers taking into account processes occurring on the Sun which cause amplification in the solar radiation, and, additionally, cause perturbations in the SW with a bulk of particles.

This is, particularly, related with so called 'space weather'. Let us recall the definition of the space weather recently given by Lilensten and Belehaki (2009), that considers the physical and phenomenological state of natural space environments around the Earth. This aims, through observation, monitoring, analysis and modelling, at understanding and predicting the state of the Sun, the interplanetary and planetary environments, and the Solar and non-solar driven perturbations that affect them, also forecasting and, now, casting the potential impacts on biological and technological systems. Space weather research addresses physical processes in space ranging from solar phenomena to their impact on interplanetary space, geo-space and at the surface of the Earth.

First of all, Sun has a huge influence on the Earth's atmosphere and biosphere. The Sun–Earth connection has been investigated since the 17th century and nowadays a number of effects are already known.

The solar activity (see figure 5), especially Coronal Mass Ejections (CMEs), solar flares and energetic particles are the major factors controlling the space weather which is greatly influenced by the speed and density of the SW and the interplanetary magnetic field (IMF) carried by the SW plasma. A variety of physical phenomena are associated with space weather, including geomagnetic storms, energisation of the Van Allen radiation belts, geomagnetic activity, ionospheric disturbances and scintillations, aurora, and geomagnetically induced currents in the Earth's surface (see a detailed discussion in Singh *et al.*, 2010).

However, it is not yet clear how these effects can cause particular distribution in the Earth's atmosphere.

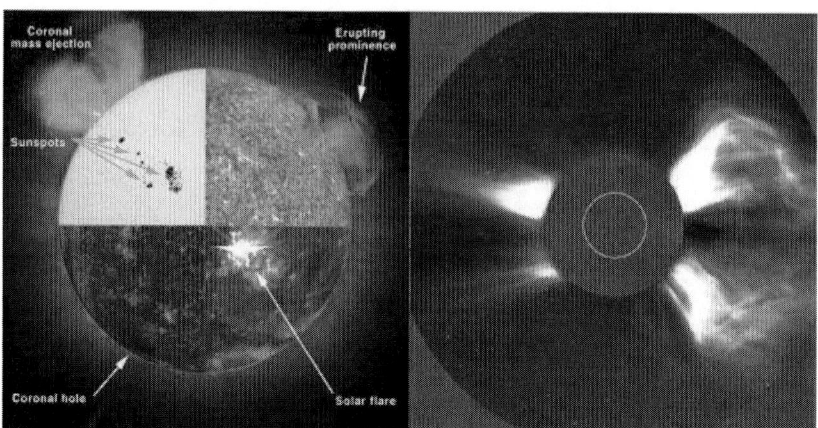

(Source: SOHO, http://sohowww.nascom.nasa.gov).

Figure 5. Left: The structures on the Sun which can influence the Earth's atmosphere and magnetosphere. Right: The huge Coronal Mass Ejection observed on November 6, 1997 by satellite SOHO. In the centre of the figure, there is a filter which is protecting the device from damage, due to which the inflow of plasma towards the Earth cannot be seen.

Let us shortly describe the basic solar structure, focusing on the structures which indicate activity of the Sun and those which can affect the solar radiation. The solar atmosphere has three parts: photosphere (visible solar surface), chromosphere and corona. As it is shown in figure 5, the structures which can be seen in the solar photosphere are *sunspots.* A sunspot represents an area on the Sun's surface (photosphere) with a strong magnetic field and lower temperature than the average magnetic field and temperature in the photosphere. During hundreds of years observations of the Sun showed that the number of sunspots varies in an 11-year cycle. At the start of a sunspot cycle the number of sunspots is at the minimum and most of them are at around 35° from the solar equator. At a solar maximum, when the sunspot number peaks about 5.5 years later, most of the sunspots are within just 5° of the solar equator. Many works tried to connect the 11-year solar activity with global or local climate changes. Among these, recently Hodges and Elsner (2011) investigated the relationship between the US hurricanes and solar activity, and found that the probability of three or more US hurricanes decreases from 40 to 20%, as sunspot number increases from lower to upper quartile amounts. Additionally, it is interesting that sunspots are located very close to the solar equatorial belt, similar as violent cyclonic motion in the Earth's atmosphere.

Prominences are bright clouds of gas forming above the sunspots in the chromospheres (the part of solar atmosphere located between photosphere and corona) which follow the magnetic field line loops. There are two types of this phenomenon. First, the so called "quiet" prominences are forming in the corona about 40,000 km above the photosphere. Sometimes they form loops of hydrogen that follows loops in the solar magnetic field. Quiet prominences can last from several days to several weeks. A second type, is the so called "surge" prominence, which is more dynamic and energetic, and last up to a few hours, shooting gas up to 300,000 km above the photosphere.

Solar flares are eruptions which are more powerful than surge prominences (for more details see Aschwanden, 2002). They last only from a few minutes to a few hours, and a lot of ionized material is ejected in a single flare. Unlike the material in prominences, the solar flare material moves with such an energy that is high enough to escape the Sun's gravity. Particles created in a flare with enough energy to escape the Sun's gravity can reach Earth and interact with the geomagnetic field. Large flares are often associated with huge ejections of mass from the Sun. These CMEs are balloon-shaped bursts which are rising above the solar corona, expanding as they climb (see figure 5, right).

Solar plasma is heated to tens of millions of degrees, and electrons, protons, and heavy nuclei are accelerated to high velocities reaching the speed of light. CMEs occur in multi-polar topologies in which reconnection between a sheared arcade and neighbouring flux systems triggers the eruption. This reconnection removes the non-sheared field above the low-lying, sheared core flux near the neutral line, thereby allowing this core flux to burst open (see numerical simulations in Antiochos, 1999; and a review of simulation in Jacobs and Poedts, 2011).

Solar flares are known to contain as much as 10^{29} J of energy and can accelerate electrons and protons to energies of several MeV and even hundreds of MeV. As we mentioned above, in a CME particles are high accelerated, i.e. CME events, as they propagate away from the Sun, are also capable of accelerating interplanetary particles to higher energies. The relationship of these CME events to solar phenomena, such as sunspots and flares, is not yet well understood. However, CMEs are now known to be important sources of disturbances of the interplanetary medium and of the space environment of the Earth, even during years of solar minimum, since they appear on average once per day in the minimum of solar activity.

The corona is very hot with temperatures that can reach 3.6 million K. Such high temperature causes high level of ionization of elements (such as atoms like iron which can have 9 to 13 electrons ejected, i.e. can be the 13th time ionized) in the corona, and also high kinetic energy of particles. Most of the corona is trapped close to the Sun by loops of magnetic field lines. In X-rays, those regions appear bright. Some magnetic field lines do not loop back to the Sun and will appear dark in X-rays. These places are called coronal holes.

Particles from the corona also stream out along the magnetic field lines of the Sun that extend into interstellar space. Fast-moving particles can escape the Sun's gravitational attraction.

Moving outward at hundreds of km/s, these positive and negative charges travel to the farthest reaches of the solar system. They are called *the solar wind* (more details for modelling of the solar wind can be found in Cohen, 2007). The wind consists primarily of electrons and protons, but also alpha particles (He nuclei) and other ionized species. The SW particles passing close to a planet with a magnetic field are deflected around the planet. Fluctuations in the solar wind can provide energy to the trapped charged particles in the planet's radiation belts.

Particles having enough energy can leave the belts and spiral down to the atmosphere to collide with molecules and atoms in the thermosphere of the planet. In particular, interaction of the SW with the Earth is very interesting, which are interactions of the SW disturbances with geo-space (Watermann *et al*, 2009).

There are two types of the SW: i) fast wind with velocity around 800 km/s (at the Earth) and ii) slow wind with the velocity around 400 km/s (at the Earth). The fast wind is observed over coronal holes, that is, low density regions in the corona. The slow wind is associated with coronal streamers which straddle regions of different magnetic polarity.

2.3.1. Solar and Geomagnetic Activities

The geomagnetic field shows different values over the globe: as it can be seen in figure 6, there are places with weaker geomagnetic field, and there are some anomalies in the magnetic field. This is the well known, so called, South Atlantic Magnetic Anomaly (SAMA). The geomagnetic field is affected by the particles which are coming from the Sun, especially in the regions of geomagnetic anomalies (Abdu *et al.*, 2005). Corpuscular emissions from the Sun are able to enter the Earth's upper atmosphere, especially at high latitudes and can bring about significant heating of the lower thermosphere. Chapman and Ferraro (1931) were the first ones who proposed the existence of a stream of ionized particles, ejected from the Sun and interacting with the Earth's atmosphere.

Considering this effect they explained the occurrence of aurora and other related disturbances. Most of the plasma that enters the Earth's magnetosphere is directed along magnetic field lines into the polar regions, where it is deposited in the ionosphere and subsequently the thermosphere.

Moreover, the SW affects the shape of the geomagnetic field, forming magnetopause, which is the boundary where the reduced SW dynamic pressure is equal to the magnetic pressure of the Earth's outer magnetosphere.

The geomagnetic activity is closely related to the solar activity, and there are four classes of activity (Legrand and Simon, 1989): i) the magnetic quiet activity due to the slow SW flowing around the magnetosphere; ii) the recurrent activity due to high speed SW; iii) the fluctuating activity related to the fluctuating SW, and iv) the shock activity due to the shock events caused by violent processes on the Sun (such as CME).

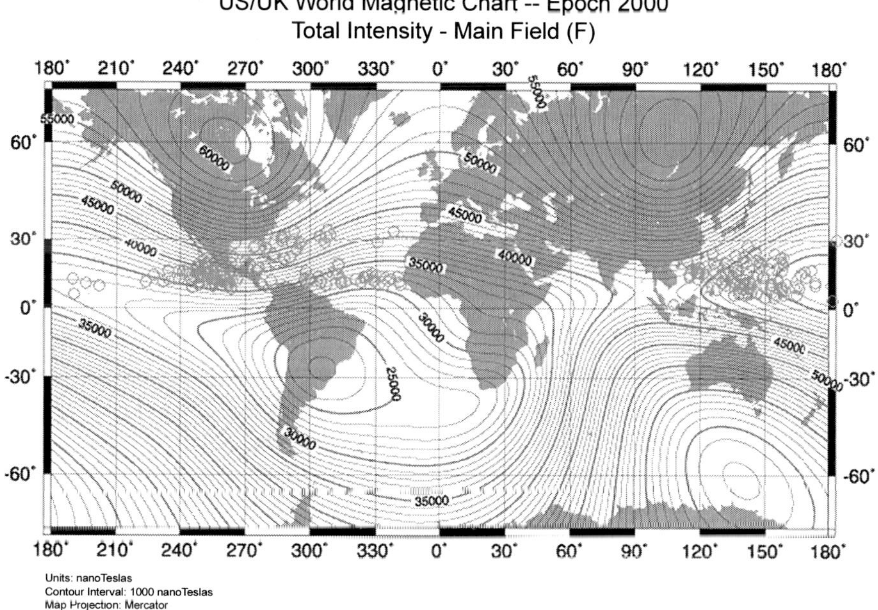

The used data for hurricanes are the same as in Figure 1.

Figure 6. World magnetic chart for Epoch 2000 (source: National Geophycal Data Center, http://www.ngdc.noaa.gov). Magnetic field total intensity distribution, represented by iso intensity lines, over the globe, with the green points the places of hurricanes in the period 1997-2005 are shown.

It is not yet clear how the perturbation in the geomagnetic field can reflect to some perturbations in the Earth's atmosphere responsible for cyclonic motion, but there are several indications. Let us recall several investigations which indicate influence of changes in geomagnetic field to atmospheric changes. It is interesting that high correlation coefficients were found between geomagnetic activity, the sea level atmospheric pressure and the surface air temperature (Bucha and Bucha, 1998). In Bochnícek (1999) the tropospheric temperature and pressure fields on the Northern Hemisphere in the winter periods 1952-1996 were investigated and they found that, in composite maps of those fields created for the high and low geomagnetic activity, phases show clear differences between different levels of geomagnetic activity. Veretenenko *et al.* (2005) noticed the coincidence of the intensity in the solar and geomagnetic activities and the changes in the air pressure intensity in the northern Atlantic. They stated that weakening the process of cyclogenesis (pressure increase) in the northern Atlantic seemed to coincide with the increase in the solar and geomagnetic activities, while the strengthening of the cyclogenesis processes (pressure decrease) was noticed during the period of the sunspot intensity weakening. Citing previous studies, the authors concluded that the impact of the solar activity on the development of cyclones was noticeable to the decimal level. To find the relationship between hurricanes and cosmophysical phenomena series in time frequency space, Perez-Pereza *et al.* (2008) used the Wavelet Coherence analysis and found that the more prominent periodicity with a coherence > 0.95 was of 30±2 years. In that context, Christoforou and Hameed (1997) arrived to interesting results (but for the Pacific action centres). Analyzing the relationship between the solar activity (represented by the annual number of sunspots) and action centres in the Pacific, they found a link between the extreme values of the solar activity and the position of these action centres. During the minimal solar activity, the Aleutian minimum moves towards the east, and the Hawaiian maximum towards the south. Since changes in the action centres influence the paths of the cyclones, considerable exceptions to the values of the climate elements on the regional level can actually be induced by extreme phases in the solar cycle. Besides statistical significance of the relationship between the solar activity and action centres, their results have practical significance in finding the mechanism of the connection and prediction of the changes of climate elements.

It should be noted that Marhavilas *et al.* (2004) have pointed out that the space weather influence on the Earth's weather and climate is still a developing topic. According to Radovanović and Gomes (2009), the locations of magnetospheric doors change during magnetic storms and strong SW. The degree of their opening is controlled by geomagnetic field, while the place and time of opening is determined by inter-planetary magnetic field (IMF). The SW kinetic energy determines which magnetospheric latitude the particles will reach.

Lilensten and Bornarel (2006) have emphasized that we are partly capable of describing the solar magnetic field but quite incapable of predicting it, with its various irregularities and, in particular, the triggering of coronal mass ejections. The same can be said of the photon flux and of life on the earth, in particular in the ultraviolet and X-rays. In the interplanetary medium, we cannot quantify the dynamic pressure of the SW or the frozen interplanetary magnetic field found there. Consequently, it is as yet impossible to determine in advance the position of the magnetic shield formed by the magnetopause: is it on this side or the other of the orbit of geostationary satellites? The characteristics of the radiation belts are not yet well known either. Furthermore, they also depend on the cosmic radiation. The phenomena which enable solar particles to enter the magnetosphere are still not understood: the aperture on the day side when the solar magnetic field reverses is only a mere model, a theory which stands up better than others to the facts. Our knowledge concerning the porosity of the magnetospheric wall or the collisions in the reconnection zone on the night side is relatively poor, for lack of observations.

In figure 6, the places of hurricane occurrence in period 1997-2005 on the magnetic map were plotted. As can be seen in this figure, most of the events were starting in the regions with similar values of the total intensity of the main magnetic field.

3. Some Possible Influence of the SW on the Formation of Cyclones

It is well known that the SW affects upper layers of the Earth's atmosphere. There is evidence that dynamic processes are enhancing the atmospheric response to solar irradiance variability (Labitzke and Van Loon 1988). Additionally, there are indications that the SW is related to the cloud cover (Pudovkin and Veretenenko, 1995 ; Svensmark and Friis-Christensen, 1997). Moreover, the SW may induce changes in the global atmosphere-ionosphere circuit that would affect the rate of initial ice nucleation via electro-scavenging (Tinsley and Heelis, 1993). Some modelling of the SW interaction show that high particles from the SW are able to perturb the stratosphere, significantly the winter stratosphere (Arnold and Robinson, 2001). It seems that there is also a connection between atmospheric temperature and solar irradiance - geomagnetic activity (Lu *et al.*, 2007).

The main question is: can the SW affect the cyclonic motion, or is there any possibility that violent cyclonic motions are caused by particles from the SW?

Possible influence of the SW on the formation of cyclones in the atmosphere is developed on the assumption that the particles from the SW penetrate through the Earth's atmosphere and seize air masses by their hydrodynamic pressure, thus making winds (Stevančević *et al.*, 2006). The inflow of charged particles (protons, electrons and nucleons) is necessary for the process to occur. Therefore, according to the mentioned hypothesis, so that the cyclogenesis occurs anywhere on Earth, the emission of corpuscular energy from the SW should exist and can be registered immediately before that.

The question, now, is how the bulk of particles from the SW can penetrate deeply in the Earth's atmosphere and which is the way that circular motion is starting. There are two possibilities for penetration of the geomagnetic field: the SW penetration through planetary magnetic field around the North and South poles (figure 7) and particles penetrate through geomagnetic anomalies (Gomes and Radovanović, 2008).

In both cases (above north and south poles), the magnetic reconnection observed in turbulent environment in the SW near the Earth (Retinò et al. 2007) has an important role. Magnetic reconnection is a topological rearrangement of magnetic field that converts magnetic energy to plasma energy. This is a universal process in space that plays a key role in various astrophysical phenomena such as star formation, solar explosions or the entry of solar material within the Earth's environment (Zweibel and Yamada 2009). Especially, in the SW close to the Earth, it causes the accretion of charged particles, and can be directly shot in the Earth's atmosphere and therefore, it may cause that fast particles from the SW can penetrate deeper in the atmosphere of the Earth.

The penetration of the SW charged particles through geomagnetic anomalies was considered by Gomes and Radovanović (2008). The term tropical areas should be taken conditionally because, with a sharper angle of incidence, the SW particles can be transported considerably deeper in the area of moderate and even sub-polar latitudes (figure 8). The authors consider that in dependence on electromagnetic and physical-chemical characteristics of the

SW, the intensity of disturbances in the atmosphere is also dependent. Moreover, the SW has a distribution along longitudes and latitudes. As an example, in figure 8, the time series of the solar wind velocity (1h averaged) obtained from two satellites ACE (blue line) and Ulysses (red line), can be seen.

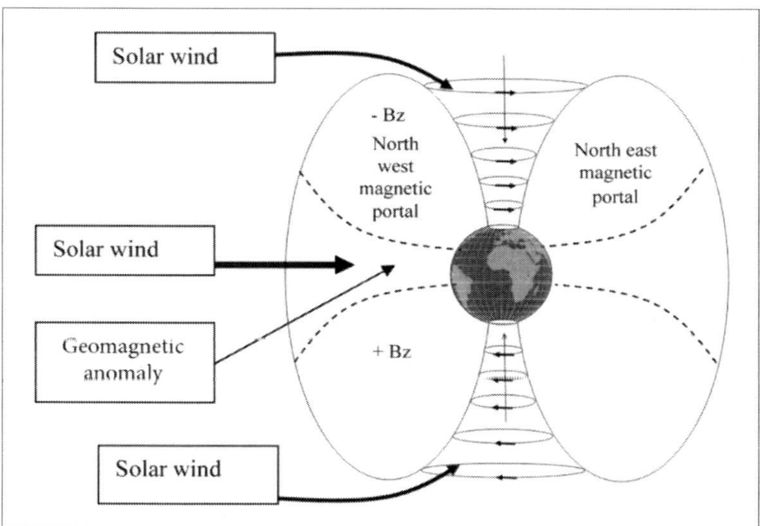

Figure 7. Schematic representation of the penetration of particles from the solar wind towards the Earth.

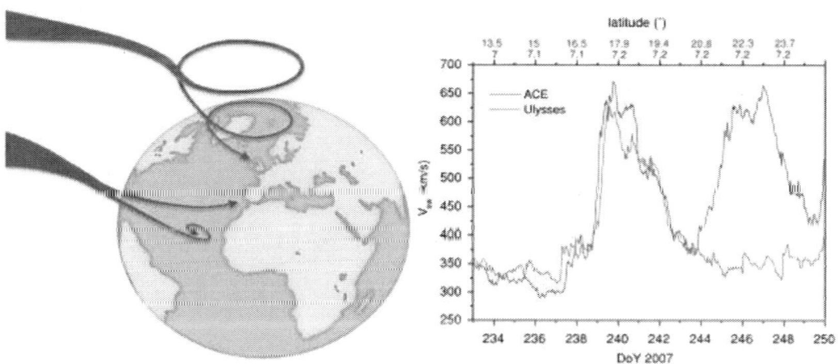

Figure 8. Left: Schematic representation of the way of SW penetration towards topographic surface (Gomes et al., 2009); Right: The difference in the velocity of SW observed by ACE (blue line) and Ulysses (red line) in August 2007 at different latitudes (D'Amicis et al., 2010).

According to Stevančević (2009), each stream of particles has its own separate shell which does not allow the mixing of individual internal streams. Moreover, it can be seen in figure 9 that there is no circular movement in the centre of a jet stream and only the radial motion in a straight line occurs under the influence of kinetic energy. If the assumed causality proves justified, it can be said that the internal wall at cyclones is the consequence of the influence of centrifugal force and the law on the circulation of the vector of the magnetic induction.

Analogically observed, the internal wall of cyclone is on the radius where centrifugal force and the force of circulation of the vector of the magnetic induction are equalized.

If one considers that the SW represents a number of charged particles, then we can say that it is acting as a current and that the density of electric convectional current is homogeneous. The lines of the magnetic field can be taken as concentric circles in the levels which are perpendicular to the axis of the stream. In the interior of the stream, the circulation of the vector *B* of IMF on spherical contour (figure 9), the radius *r* which is smaller than *ra*, is equal with the convectional current of the SW particles which pass through that contour.

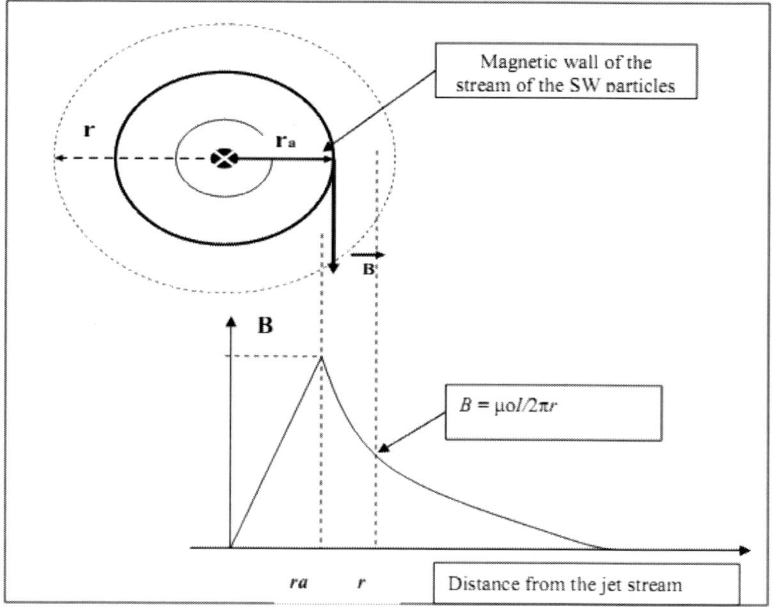

Figure 9. Graph representation of air mass motion within the solar wind stream.

The intensity of the magnetic induction linearly grows with the increase of the radius of the contour in the interior of the stream through which the SW charged particles move. The stream of the SW particles gets the magnetic shell and becomes an ordinary conductor of electric convectional current that is coming from the Sun. Moreover, it can be noticed that the vector of the magnetic induction is the largest on the walls of the stream, which points out that the SW particles cannot get out of it. On the basis of the law on the circulation of the vector of the magnetic field, the motion of electrically charged particles of the SW can simultaneously be radial and circular.

When the two previous considerations are applied on charged particles from the SW which are moving in geomagnetic field, the electromagnetic force ($F = qv \times B$, where q is the electric load of particles, and **v** is the speed of electrons) will force a circular motion of the bulk of electrons.

In both, sub-polar and tropical regions, the results is the reduction of kinetic energy after the penetration into deeper layers of the troposphere due to friction with denser layers of air and, by itself, to the weakening of the power of the magnetic shell of the main stream of the SW.

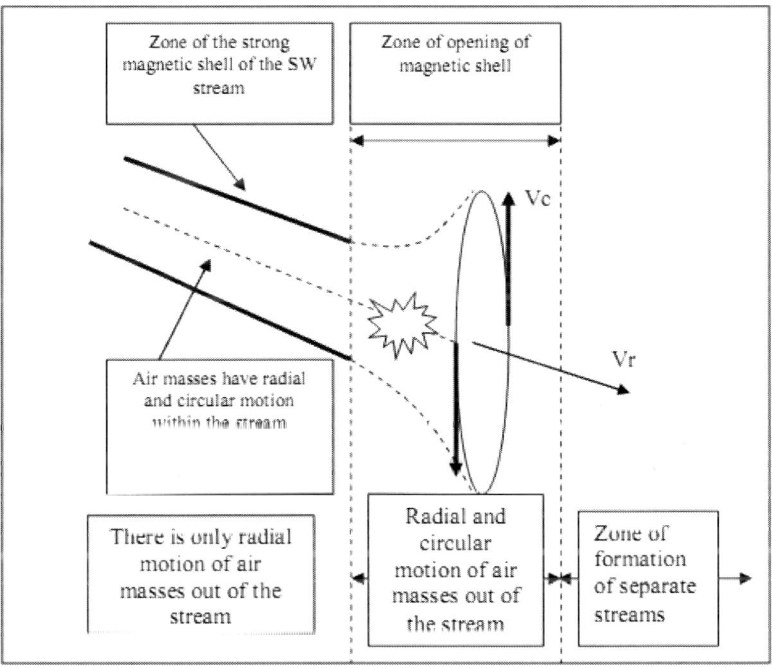

Figure 10. Schematic representation of the opening of magnetic shell of the main SW stream.

Moreover, depending on the parameters of the SW it depends on a number of smaller streams which will appear at lower heights (figure 10).

Stevančević (2009), Radovanović (2010), Nikolić *et al.* (2010) consider that free electric loads enter the atmosphere and change the existing synoptic situation in the zone of opening of jet field. In that zone, it results into the separation of some streams of the SW which continue penetration towards lower layers of the troposphere under different angles.

Let us assume that by entering into denser layers of the atmosphere, the jet stream of the SW particles seizes air masses and makes "earthly" winds. Such assumption is burdened with the lack of detailed parameterization that would be presented by the corresponding model.

It may be that, after opening the magnetic shell, i.e. opening the jet field, protons and electrons will separate from the main stream to the opposite sides due to opposite polarities (figure 11). Holding the circular speed they had in the jet field, electrons in the northern hemisphere turn right, while protons turn left from the direction of the radial speed of particles. Gravitational force represents special factor which influences the motion of free electric loads towards the ground.

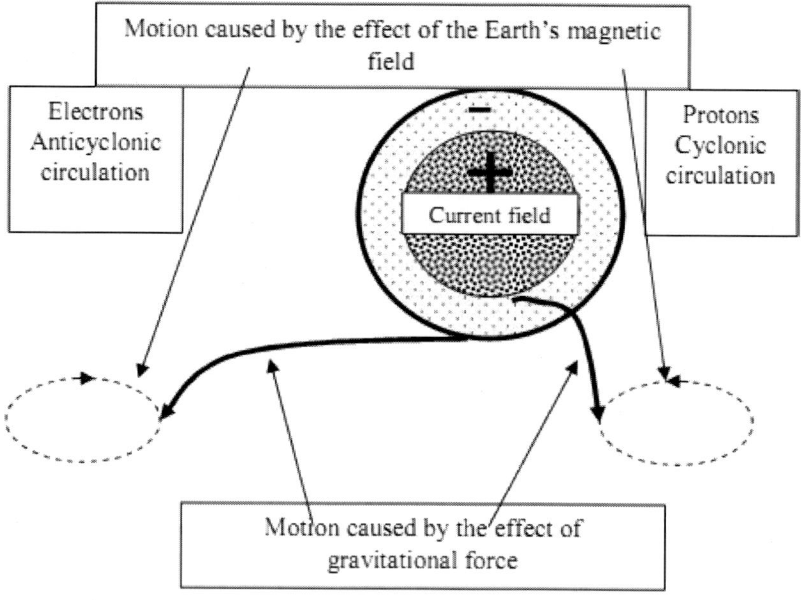

Figure 11. Schematic representation of the separation of protons and electrons from the main stream of the solar wind in the geomagnetic field.

3.1. Theoretical Approach of Particle Penetration in the Earth's Atmosphere

Taking the basic electromagnetic equations one can roughly estimate kinematical parameters of penetrating cloud, assuming that SW has the electric load q, mass m and speed **v** at the time of contact with the Earth's atmosphere. At this place the Earth's magnetic induction is *B*. We will now consider a very simple case taking the relation for the electromagnetic force:

$$\mathbf{F} = q\mathrm{v} \times \mathbf{B},$$

Since even a simple relation can give us the estimate of the radius (r) of trajectory of a bulk of charged particle as:

$$mv^2/r = qvB\sin(\boldsymbol{\theta}\mathbf{sw}), \Rightarrow r = mv/\ qB\sin(\boldsymbol{\theta}\mathbf{sw}),$$

where θsw is the angle between direction of motion of SW and magnetic field.

From this very simple relation, one can conclude that the radius is larger for higher velocity which is penetrating a weaker magnetic field. To penetrate deeply in the atmosphere, the optimal values of the SW velocity are needed, as well as the orientation and charge of corresponding magnetic field. In the case of small r there will be present a well known effect that particles from the SW are drifting around geomagnetic lines, and for larger r (as e.g. larger than Earth's radius) the particles penetrate without circular motions in the Earth's atmosphere.

However, in the case of an optimal ratio of the velocity from one side and the current of the SW and magnetic field at the place from other side, as well taking the angle between the direction of the SW and magnetic field, there may be present a penetration of the SW through Earth's atmosphere as a circular moving current. In this case, one can roughly estimate (taking very simple relations) that the step of the trajectory of the penetrating particles is given by:

$$d = 2\pi\ m\ v\cos\boldsymbol{\theta}\mathbf{sw}/q\mathbf{B},$$

and the time for which a particle makes a circle is:

$$t = 2\pi r/v = 2\pi m/q\mathbf{B}.$$

Moreover, one can estimate the frequency of the circular motion as:

$$f=1/t = qB/2\pi m$$

It is interesting, that the time and frequency do not depend on velocity. On the other hand, one can expect that the velocity should be changing during travelling of the charged cloud in the Earth's atmosphere. Namely, the speed v is gradually reduced (which means: the particles are slowing down), and, consequently, the radius r becomes smaller as well as the step. For this reason, in one moment (in one layer of the Earth's atmosphere) the penetration of the charged cloud will be almost stopped, and almost circular motion of the charged cloud will be present. This may cause, or contribute, to cyclonic motion (Radovanovic *et al.*, 2003).

Based on the considerations presented above, it comes out that, in certain synoptic conditions the separate streams of the SW may make relatively small locations with cyclonic movement of air masses. Several relatively small locations with low air pressure can be noticed over the northern Atlantic and north-western Europe in figure 12.

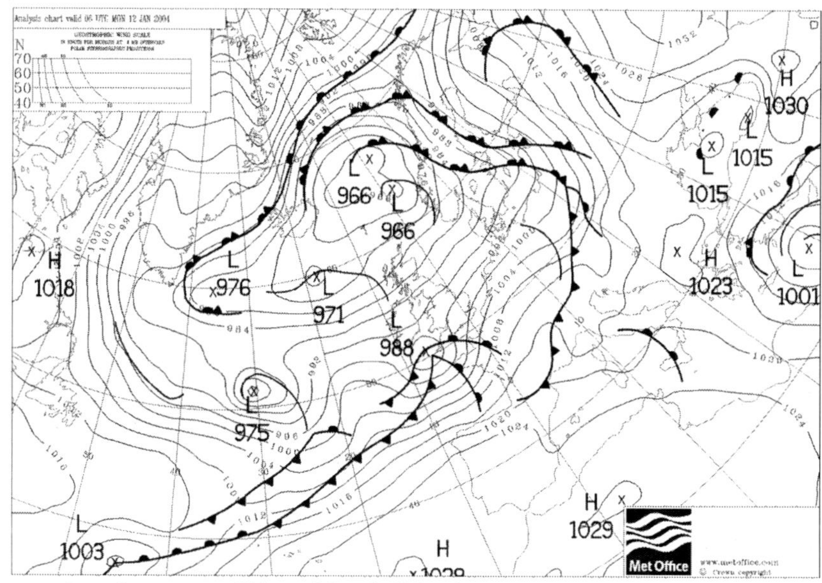

(Source: http://meteonet.nl/aktueel/brackall.htm).

Figure 12. Synoptic situation on January 12th 2004 several hours after the tornado in Ireland.

In any case, one can conclude that the amplified (turbulent) SW in combination with the process of reconnection may accelerate the charged particles and shoot them down deeply in the Earth's atmosphere. The bulk of charged particles can perturb an air mass in a way that cyclonic activity is starting.

3.2. Some Particular Cases of Violent Cyclonic Motions and Their Connection with the Solar Wind

Here, we are going to present several particular cases where the appearance of violent cyclonic motion may be connected with the SW. First, let us discuss the unusual case of tornado in Ireland and after that hurricanes Katrina, Rita and Wilma.

3.2.1. Tornado in Ireland

Figure 13 shows that CH75 coronary hole and energetic source 10536 were in geo-effective position on the Sun the day before tornado in Ireland. Throughout the morning hours (up to 10:00 UTC), the speed of the SW was reaching 723 km/s (figure 14).

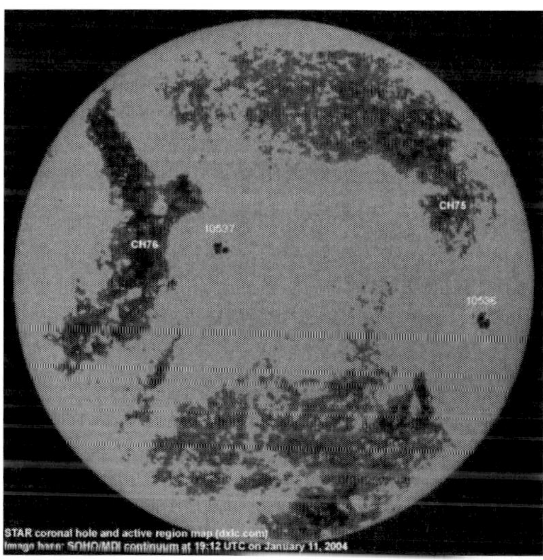

(Source: http://www.dxlc.com/solar/index.html).

Figure 13. Position of coronary holes and energetic sources on the sun one day before the phenomenon of tornado in Ireland.

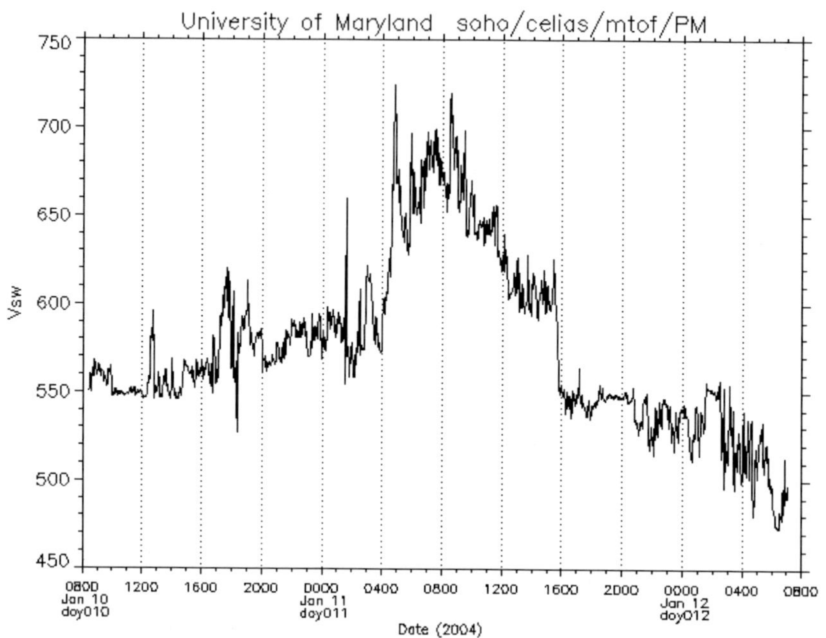

(Source: http://umtof.umd.edu/pm/crn/).

Figure 14. Speeds of protons immediately before the phenomenon of tornado in Ireland.

It can be noticed from figures 15 and 16 that the flux of particles had the maximum in all energetic ranges on January 10th 2004. In figures 17 and 18, the temporal sequence of events can be noticed immediately before and after the phenomenon of the mentioned tornado (Mukherjee and Radovanović, 2011). The origin and development of tornado in any part of Europe, i.e. northern hemisphere in the winter period, represents unusual and rare phenomenon.

In Ireland, it occurs averagely less than 5 times during January (Tyrrell, 2007). The mentioned author emphasized that the tornado which occurred on January 12th 2004 around 02:00 UTC, lasted about 20 minutes according to the statements of witnesses. The trajectory of the tornado was about 4 km long. Quantitative parameters that would show the development of synoptic situations in details were most often without relative indications. The reason lies in the fact that the trajectories of tornadoes sometimes seize relatively small areas which are, almost by the rule, far away enough from meteorological stations by which the intensive disturbances of the local character would be detected.

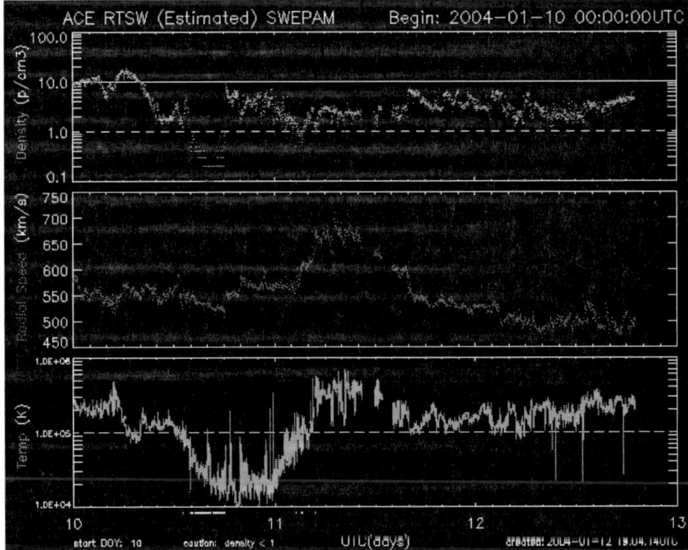

(Source: http://www.swpc.noaa.gov/ace/SWEPAM_7d.html).

Figure 15. Densities, speeds and temperature of protons show sudden rise one day before the phenomenon of tornado in Ireland.

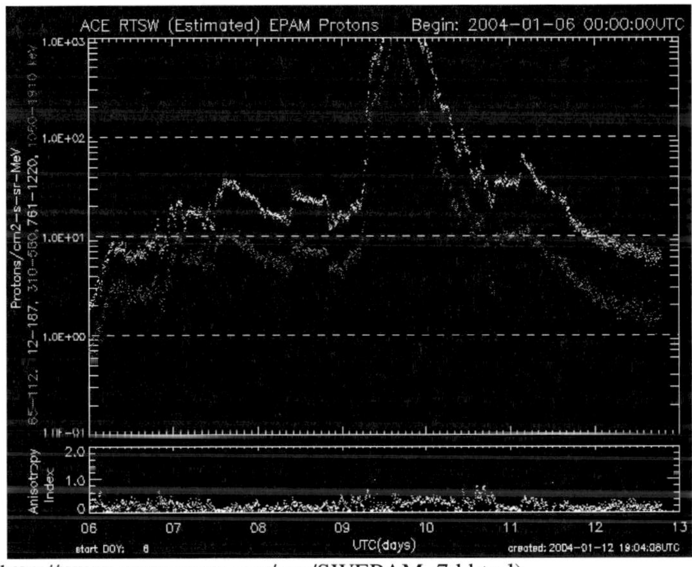

(Source: http://www.swpc.noaa.gov/ace/SWEPAM_7d.html).

Figure 16. Flux of protons in corresponding energetic ranges immediately before the phenomenon of tornado in Ireland.

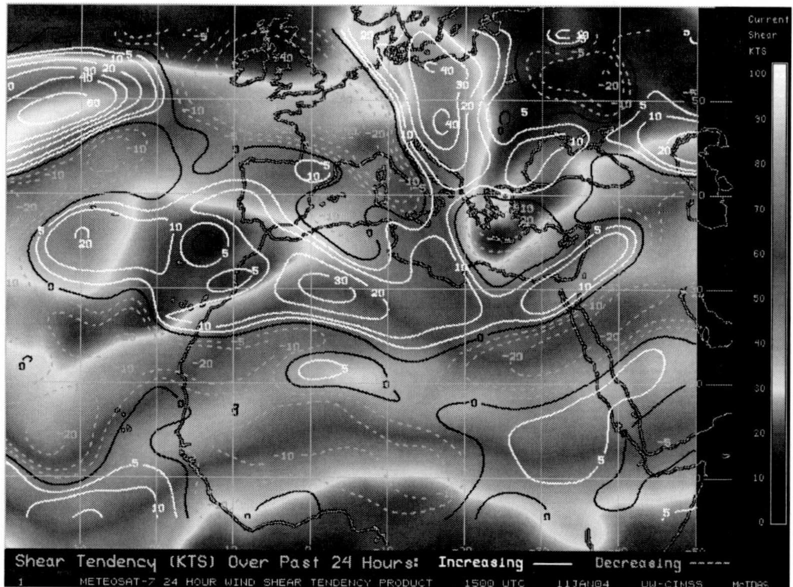

(Source: http://cimss.ssec.wisc.edu/tropic/real-time/europe/winds/wm7sht.html).

Figure 17. Approximately 9 hours before the phenomenon of tornado in Ireland, strong stream of wind had existed, directed towards the northeast with speeds over 80 knt.

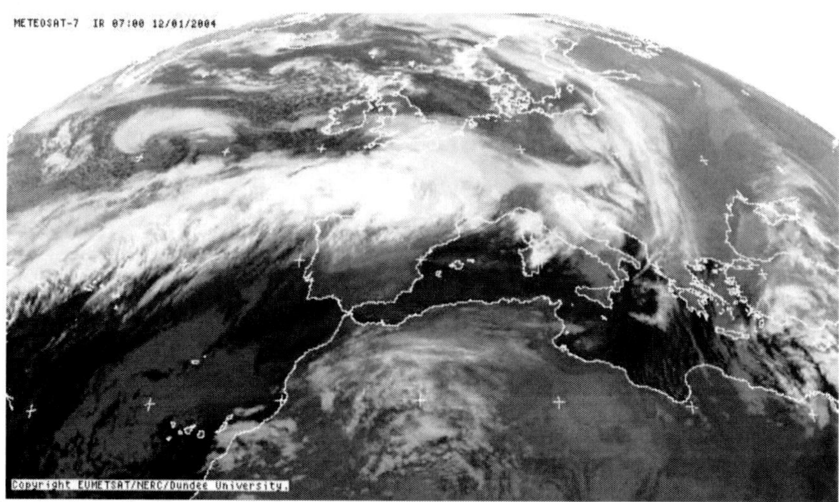

(Source: http://www.sat.dundee.ac.uk/pdus.html).

Figure 18. Satellite image of air mass motion over Western Europe on January 12th 2004, 5 hours after the phenomenon of tornado in Ireland.

As tornadoes only affect a small area, the probability of them being observed at a meteorological station is very small (Leitão, 2003).

In cases when alternating sudden increase and decrease of the flux of corpuscular energy is present, it seems that it comes to the pulsation in the motion of cyclones. According to Veretenenko, Thejll (2004), the weather chart analysis showed that the deepening of the cyclones correlated with solar proton events (SPE) under study may be considered as the cyclone regeneration. Indeed, the majority of these cyclones are formed near the eastern coasts of North America and, when they travel near Greenland, they have already reached their maximum development. However, the observed intensification of their regeneration suggests that energetic SPE seem to create conditions contributing to this process. In the case of the tornado in Ireland, it seemed that it came to the penetration of one relatively weak stream that had limited effect both in time (around 20 minutes) and space (approximately 4 km). In theoretical sense, Schielicke and Nevir (2009) also emphasized the potential possibility of the influence from outside. If intensity is expressed as lifetime minimum pressure, the theoretical number–intensity (pressure ratio) distributions are power law distributed. This cannot be proved yet. Power laws often describe open systems with external forcing and dissipation that organize in a critical, non-equilibrium state. In the case of atmospheric depressions, the imbalance is represented by the disturbance of the hydrostatic balance.

3.2.2. Hurricanes Katrina, Rita and Wilma

Let us assume that the penetration of the SW over the Atlantic geomagnetic anomaly caused the phenomenon of hurricane Katrina on August 23rd 2005. Similar to the tornado which was mentioned in the case of Ireland, sudden influx of protons in certain energetic ranges had preceded the cyclogenesis, as can be noted in table 1.

On the basis of table 1, the number of charged particles per unit of surface increased in all energetic ranges up to August 23rd and 24th 2005. After that, the values were decreasing, but they were still with considerably higher values than before the phenomenon of hurricane, except the protons in the range > 100 MeV. Tropical Depression Twelve was formed over the south-eastern Bahamas at 21:00 UTC on August 23, 2005. As the atmospheric conditions surrounding Tropical Depression Twelve were favourable for tropical development, the system began to intensify and was upgraded to Tropical Storm Katrina on the morning of August 24 (Knab *et al.*, 2006). It can be noticed from table 1 that the first more significant influx occurred on August 22nd at protons in the range of 1 and 10 MeV.

Table 1. Number of protons of certain energetic ranges several days before and after the phenomenon of hurricane Katrina

	(protons/cm ^{2}-day-sr)		
Date	>1 MeV	>10 MeV	>100 MeV
2005.08.20	1.1e+06	1.6e+04	4.0e+03
2005.08.21	1.1e+06	1.6e+04	4.3e+03
2005.08.22	1.0e+07	7.2e+05	4.8e+03
2005.08.23	1.4e+08	1.7e+07	1.1e+04
2005.08.24	2.6e+08	5.1e+06	4.8e+03
2005.08.25	3.2e+07	2.9e+05	3.2e+03
2005.08.26	2.7e+06	4.6e+04	3.6e+03
2005.08.27	2.3e+06	2.2e+04	3.3e+03

Source: http://umtof.umd.edu/pm/crn/.

On August 24th 2005, the penetration of air masses towards north-western Europe occurred. The main stream of the SW over the Atlantic anomaly divided in two smaller streams, one which caused hurricane Katrina and the second one which caused cyclone that moved from the west towards the east, i.e. northeast. The following days, it resulted into the intensification of cyclones both north of England and in the western Atlantic (figure 19).

(Source: http://www.sat.dundee.ac.uk/pdus.html).

Figure 19. Satellite image of air mass motion over Western Europe on August 24th 2005.

Considering the mentioned idea, the angle of incidence of the SW towards the ground was considerably higher at hurricane Katrina than at the stream that moved towards Europe. However, it seems that individual separations from the main stream were not just connected with these two cases. Knabb *et al.* (2006) have pointed out that a total of 43 reported tornadoes were spawned by Katrina. One tornado was reported in the Florida Keys on the morning of 26 August. On 29-30 August, 20 tornadoes were reported in Georgia, 11 in Alabama, and 11 in Mississippi. The Georgia tornadoes were the highest record in that state for a single day in the month of August, and one of them caused the only August tornado fatality on record in Georgia.

Having in mind that the energetic sources on the Sun (S583 and 10797) had preceded the mentioned processes in the atmosphere it could be assumed that a similar phenomenon was about to occur in the following rotation of the sun. That, actually, did happen. Namely, hurricane Rita appeared as TC on September 18 at 18:00 UTC after which it came to the weakening of speed. The next day, on September 19 at 18:00 UTC, it became stronger again and reached the level of TC (http://www.nhc.noaa.gov/pdf/TCR-AL182005_Rita.pdf). The temporal difference in appearance of TC Katrina and regenerated TC Rita was 26.5 days (note here that solar rotation is differential, and the period is from 25 days on solar equator to 27 days close the solar poles). The assumption that separate cyclonic movements of air masses were to appear, in this case, was confirmed again. Knabb, Brown *et al.*, (2006) have stated that at least 90 tornadoes were reported in association with Rita, mainly to the north and east of the circulation centre in portions of Alabama, Mississippi, Louisiana, and Arkansas. Rita produced the most tornadoes (56) in a single event (of 48h or less in duration) ever recorded in the area of responsibility of the Jackson, Mississippi NWS forecast office (which also includes portions of north-eastern Louisiana and extreme south-eastern Arkansas). Eleven tornadoes were reported in other portions of Arkansas, and 23 tornadoes were reported in Alabama.

According to Stevančević (2009), the number of tornadoes shows how many separate jet streams the grouped stream of the SW particles, which created hurricane, was composed. The breakdown of the primary jet stream was the consequence of the transition of the hurricane from ocean to land when it came to the sudden increase in geomagnetic induction above the land.

In the next rotation of the sun, after approximately the same temporal distance, hurricane Wilma appeared. Dvorak classifications were initiated on October 15. The system continued to organize, with the National Hurricane Centre remarking the system could ultimately become a hurricane. In contrast to previous two cases, considerably smaller number of tornadoes appeared at Wilma. Wilma produced 10 tornadoes over the Florida peninsula on 23-24 October: one each in Collier, Hardee, Highlands, Indian River, Okeechobee, and Polk Counties, and four in Brevard County (Pasch *et al.*, 2006). Data analysis, which was done using the same source as for table 1 for corresponding periods, shows similar rises, such as sudden influxes of energy, both in the case of Rita and Wilma.

Some more indications are in favor of the heliocentric hypothesis on the causality between the processes on the Sun and violent cyclonic activity. Namely, extremely low temperatures were recorded at all three hurricanes on surface of 700 mb in zone of clouds. In the case of hurricane Katrina GOES-12 10.7 μm IR images revealed cloud top brightness temperatures as cold as-87°C (http://cimss.ssec.wisc.edu/goes/blog/archives/date/2005/10). The question is how such low air temperature is formed when it is known that such extreme values cannot be found even in Siberia. Moreover, in the case of Tropical Storm 07W (Molave) east of the Philippines in the West Pacific Ocean on 16 July 2009, -92.4° C was measured (http://cimss.ssec.wisc.edu/goes/blog/archives/2993). Considering that such low values can be found even only in the upper border of the mesosphere in the environment of the earth, the question of justification of the explanation is actualized on the possible seizing of air masses of the SW up to the cloud of hurricane.

Another kind of indirect indices refers to the stress caused by the stroke of the stream of the SW into the planet, i.e. the Earth's magnetic field. Analyzing the link between geomagnetic disturbances and hurricanes, the results have shown that the connection does exist. Thus, it appears that the average Kp index has a statistically significant relationship to the maximum intensity of the baroclinically-initiated hurricane. When Kp index values are higher, the probability of a stronger hurricane is larger (Elsner, Kavlakov, 2001). Similarly, Palamara and Bryant (2004) concluded that geomagnetic activity plays an important role in recent climate change, but that the mechanism behind this relationship still needs further clarification.

3.2.3. Statistical Analysis

Additionally, we are going to present some statistical facts concerning the solar activity and hurricanes. Daily data on the solar activity in the period from 2004 to 2007 is used in this chapter (http://www.swpc.noaa.gov/ftpmenu/warehouse.html), as well as daily data on hurricanes on the whole planet in the same period (http://cimss.ssec.wisc.edu/tropic2/tropic.php?andtestie6=1).

By checking the plot of distributions of variables which describe the solar activity, it is established that the normal distribution did not appear in any of them. Therefore, Mann-Whitman U test is used for checking the significance of difference in values of these variables on days when a certain disturbance exists in the atmosphere (hurricane, tropical cyclone, tropical storm) and on days when the disturbance does not exist (von Storch, Zwiers, 1999).

It turned out that there has been a statistically significant difference at variables 1MeV protons, 0.6MeV electrons and 2MeV electrons, which means that the solar activity, presented by these variables, was significantly higher for days when there was a disturbance in the atmosphere than it was for days on which such disturbances were not recorded (tables 2-4).

Table 2. Results of Mann-Whitney U tests

Mann-Whitney U Test									
	Rank Sum – Group 1	Rank Sum – Group 2	U	Z	p-level	Z - adjusted	p-level	Valid N – Group 1	Valid N – Group 2
>1MeV protons	974046.0	93945.0	79067.0	6.116021	0.000000	6.116674	**0.000000**	1289	172
>10MeV protons	943816.0	124175.0	109297.0	0.299485	0.764570	0.302357	0.762380	1289	172
>100MeV protons	948657.5	119333.5	104455.5	1.231035	0.218311	1.232847	0.217634	1289	172
>0.6MeV electrons	967721.0	100270.0	85392.0	4.899032	0.000001	4.899594	**0.000001**	1289	172
>2MeV electrons	966431.5	101559.5	86681.5	4.650920	0.000003	4.651197	**0.000003**	1289	172

Table 3. Medians, minimum and maximum values and standard deviations of the indices of the solar activity for days when the disturbance of the atmosphere exists

	Median	Minimum	Maximum	Std. Dev.
>1MeVi protons	1.000000E+06	55000	1.100000E+09	5.455122E+07
>10MeV protons	1.600000E+04	10000	1.100000E+08	5.797749E+06
>100MeV protons	3.600000E+03	1800	6.100000E+06	1.818233E+05
>0.6MeV electrons	1.700000E+10	230000000	1.800000E+11	2.243343E+10
>2MeV electrons	3.600000E+07	700000	9.300000E+09	4.682329E+08

Table 4. Medians, minimum and maximum values and standard deviations of indices of the solar activity for days when the disturbance of the atmosphere does not exist

	Median	Minimum	Maximum	Std.Dev.
>1MeV protons	6.000000E+05	120000	6.700000E+08	5.442959E+07
>10MeV protons	1.600000E+04	12000	2.200000E+07	1.806232E+06
>100MeV protons	3.500000E+03	1800	5.800000E+04	5.385863E+03
>0.6MeV electrons	9.750000E+09	260000000	1.100000E+11	1.793924E+10
>2MeV electrons	1.500000E+07	650000	3.000000E+09	3.385925E+08

We tried to establish whether significant differences in the values of the indices of the solar activity existed for days which preceded the origin of the disturbance in the atmosphere in relation to the day when disturbance originated, i.e. for days when certain disturbance already originated. For that purpose, the design of superposed epochs was used, as well as Vilcoxon test. The chosen level of significance, by which it is concluded that the statistically significant difference existed, is $p<0.1$.

It turned out that the statistically significant difference between some days existed only for protons. By Box Whisker diagrams (figure 20), the values of these variables were shown two days before the disturbance (signs -2 and -1), the day when the disturbance occurred (sign 0) and two days after the disturbance already originated (signs 1 and 2).

Observing 100MeV protons, the significant difference existed only between the day before the origin of the disturbance and the second day after the origin.

When 10MeV protons are observed, the significant difference existed between the second and the first day before the origin of the disturbance, as well as for the day for which to the disturbance of the atmosphere occurred and the day after that.

Observing 1MeV protons, there is a significant difference between the second day before the origin of the disturbance in the atmosphere and the following three days (i.e. the day before the origin, the day of the origin and the day after the origin of disturbance).

The analysis and the results of the present study are complementary to case studies and superposed epoch analysis of Troshichev and Janzhura (2004) and Troshichev *et al.*, (2005). These authors found that surface air temperatures in Antarctica directly responded to temporal changes of the SW.

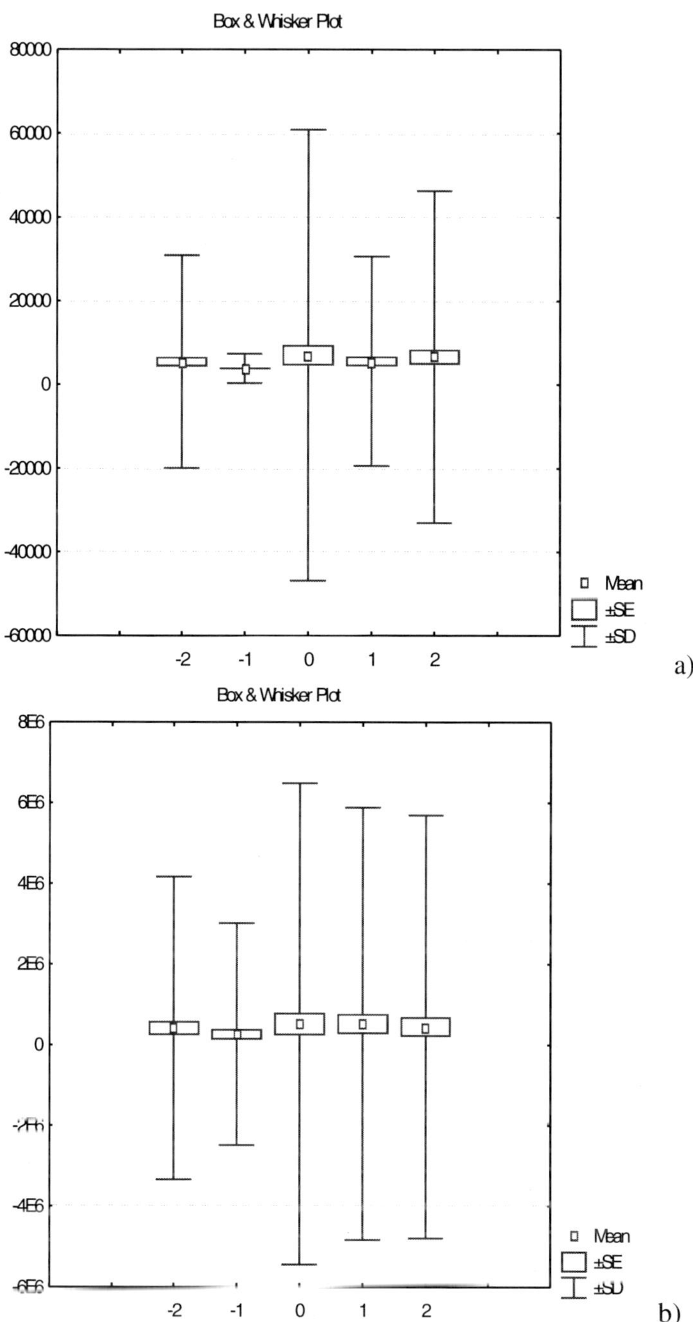

Figure 20. Continued on next page.

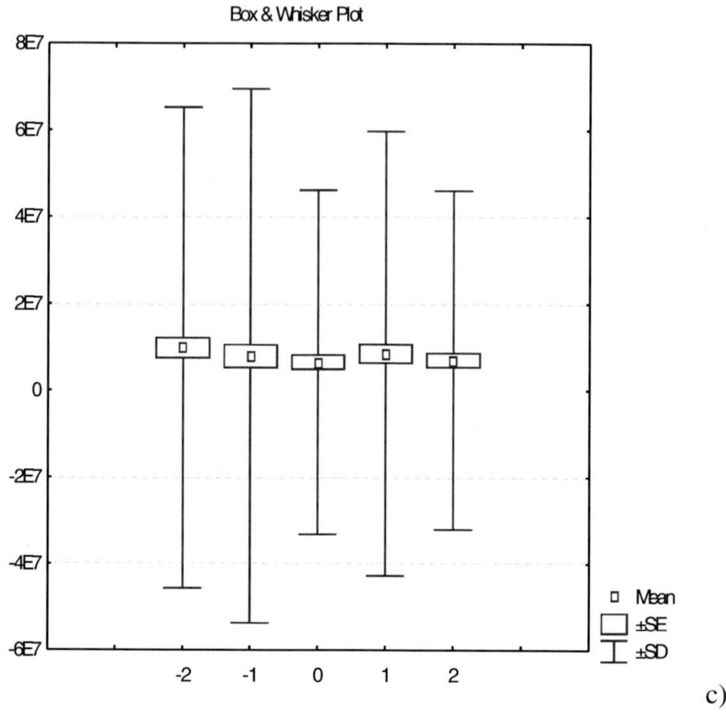

Figure 20. Box-Whisker's diagrams for 100MeV (left), 10MeV (middle) and 1MeV protons.

The SW disturbances induce changes in the atmospheric electric circuit resulting in changes of tropospheric cloud cover, atmospheric radiation budget, and dynamics. Recent observational studies by Suparta et al., (2008) and Kniveton et al., (2008) are accoridng to this interpretation.

Gabis and Troshichev (2000) concluded that the impact of short-term changes in the solar activity on baric (pressure) field perturbations is evident in the stratosphere (30 mb-level). The meridional perturbations in the stratosphere in case of the Forbush decreases and solar proton events start to develop well before the key date following growth of the UV irradiance typical of the short-term changes in solar activity. Decay of the meridional transfer occurs after the key date evidently under the influence of solar energetic corpuscular flux. Fluctuations of baric field within periods of 5 ± 10 days are typical of meridian and zonal transfer in the troposphere (500 mbar-level), intensities of meridional and zonal transfer being fluctuated oppositely in phase. The effect of the key date is not prominent in these fluctuations.

Vorticity area index, characterizing cyclonic activity in the troposphere, shows the striking correspondence to changes of the meridional transfer in the stratosphere. In order to check the relationship between the solar activity and the circulation of the atmosphere, Milovanović and Radovanović (2009) used multiple linear regressions. The obtained results have shown that the values R^2 ranged from 0.572 to 0.825, indicating the mathematical connection of the mentioned variables.

3.2.4. Short Description of Space Weather Environment - The Lunar Phases and Hurricanes

As it is well known the tidal forces could influence weather environment on the Earth. In 1964 a paper by Bradley related to the influence of tidal components in hurricane development, showed that the occurrence of heavy rainfall (10 in. and above in 24 h) is related to the lunar phase. Recently, Visvanathan (1966) published a paper, showing that such heavy falls are mostly associated with depressions, some of which intensify into cyclones and severe cyclones. So, his study was therefore undertaken to examine the dates of formation of depressions in the Indian Ocean vis-à-vis the lunar phase.

In order to examine the hypothesis of a worldwide relation between some lunar periods and tropical disturbances, Carpenter *et al.* (1972) collected first-formation dates for 1013 hurricanes and typhoons and 2418 tropical storms in both hemispheres. Using the superposed epoch method, they found a lunar synodic cycle (29.53 days) in North Atlantic hurricane and northwest Pacific typhoon formation dates. About 20 percent more hurricanes and typhoons formed near new and full moon than near the quarters during a 78 year period, showing a stronger peak at new moon than at full moon. Statistically, the existence of an effect dependent on the lunar synodic cycle is supported by a significance level of 7% on unsmoothed data from an analysis of variance for categorical data.

During the same 78 year period, North Atlantic tropical storms, that did not later become hurricanes, tended to form near the lunar quarters. Several other categories of tropical storms were not clearly related to the synodic month. Severe tropical storms in two portions of the Indian Ocean over 75 years formed more often several days after syzygy and quadrature, but this and other severe tropical storm results lack definition, probably due to poor data.

Ghayoor (1986) investigated climatic fluctuations connected with the cycles of the lunar year, using the rainfall data belonging to the Balcombe station, Sussex, England. From these data, variations in the frequency per year of daily rainfalls over 25 mm, which in this research were called daily rainbursts, were investigated, and the results show that there was a quasi-regular fluctuation, connected with the cycles of the moon, in the number of daily rainbursts.

At the Annual General Meeting 2009 of the Association of American Geographers, Peter Yaukey presented a paper where he claimed that storms that occurred in the Atlantic Ocean between 1950 and 2007 were more likely to form right after the new moon. They also intensified 49 percent more often after a new moon than at any other time in the 29.5-day lunar cycle.

Over the last century, scientific research has hinted that the moon may influence rain patterns, thunderstorms and other meteorological events. There are many explanations for this, but nothing conclusive has been shown. There are a range of possibilities. Just as the moon pulls on Earth's oceans and creates the tides, it also tugs on the air above it. Lunar atmospheric tides are thought to be weak, but could create favorable conditions for storms to strengthen.

The moon's gravity may also pull cosmic dust into the Earth's atmosphere in a cyclical fashion, perhaps seeding cloud formation and precipitation. The most promising explanation is internal tides encouraged by the lunar cycle. The currents beneath the ocean surface could circulate warm water up underneath a storm, supplying it with the energy it needs to intensify.

Only 13 percent of the world's hurricanes occur in the Atlantic Ocean. So if the moon is really influencing hurricanes, the signal should show up in the Pacific and Indian Ocean storms, too.

Therefore, we combined the data for the Atlantic, East Pacific and West Pacific from 1997 up to 2010 and calculated the statistical distributions of moon phases, under condition that time difference between appearing of hurricane and phase is less than 3 days. As it can be seen from figure 21, for all basins the largest number of occurrence of hurricanes is around the first quarter of the moon. This means that 32% (nearly 80 of 253 cases) occur nearby the first quarter, while the probability for other moon phases is somewhat lower, about 22%. However, when we combined (figure 21) the data from the Atlantic and East Pacific (due to hurricanes negative longitudes) it could be seen that for those two basins occurrence of hurricane is almost equal around new moon and the first quarter, while a slightly smaller number of occurrences is around full moon and the last quarter.

On the other hand, in the case of West Pacific (figure 21), where longitudes of hurricane occurrence are positive, almost the equal number of cases is around the first and the last quarter, while the number of occurrences around new and full moon is somewhat smaller.

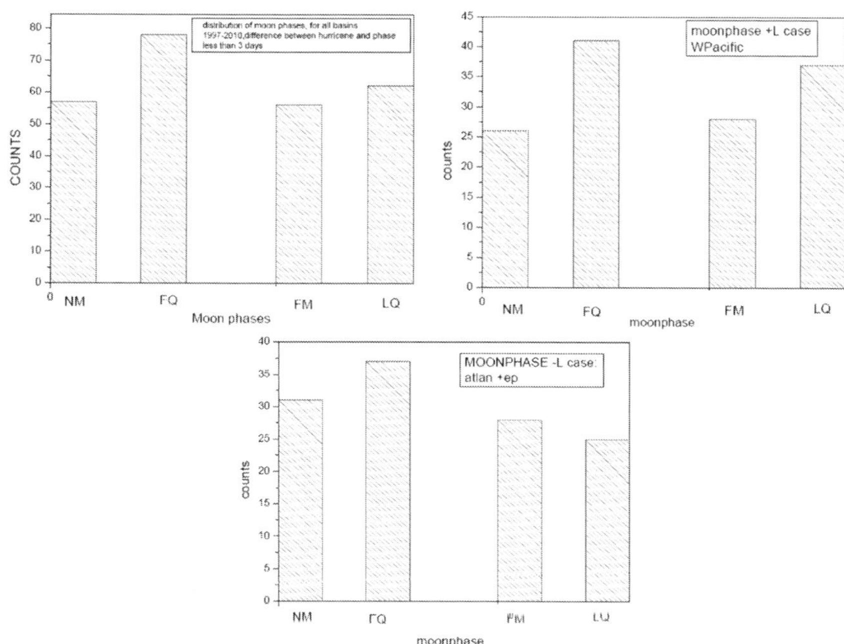

Figure 21. Distribution of moon phases within 3 days around the beginning of hurricane. Period covered from 1997 to 2010. Left panel: for the Atlantic, East Pacific and West Pacific; middle panel: West Pacific; right panel: the Atlantic and East Pacific.

In conclusion, it could be mentioned that space weather environment could be heavily disturbed by close approach of an asteroid or Near Earth Object. Actually the objects in main asteroid belt because of their masses (Kovačević, 2005; Kovačević and Kuymanoski, 2007; Kovačević, 2011) due to gravitational effects could cause such events.

Discussion and Conclusion

Here, we considered the possibility that astronomical aspects (the Earth's rotation and revolution, and the astrophysical effects of the SW) produce some special conditions which are influencing the violent cyclone motion in the Earth's atmosphere.

As it has been already seen, previous investigations on the origin of cyclones are characterized by numerous unclearness. This also happens with the astrophysical aspects of the propagation of the SW through the magnetosphere and atmosphere of the Earth and its mechanism of causing cyclone circulation. However, there are several indications (as pointed out in section 3) which are in favor that the beginning of violent cyclonic motions in the Earth's atmosphere may be caused by charged particles from the SW.

Some questions of the crucial significance for understanding the development of cyclones are still opened. Firstly, how the presented results are in accordance with the existing prognostic models, that are not containing the input parameter of the SW. Secondly, how the influx of energy is being developed, i.e. the life cycle of the cyclones depending on the rotation of the Earth throughout several days. The impression is that, similar to the wire which is winding up around the spool, the wireless inflow of energy is being developed towards cyclones. This primarily refers to the penetrations of the SW over the geomagnetic anomalies. Moreover, there is the question of the connection sun-cyclones after the appearance of the coronary holes and/or energetic sources out of the geo-effective position. This question can, partly, be explained by non linear propagation, i.e. the curving of the path of corpuscular energy from the sun (Radovanović and Gomes, 2009). Thirdly, the crucial weakness of the heliocentric hypothesis, at this moment, refers to the necessity of the parameterization i.e. mathematical modeling by which the division of the main stream into smaller streams of the SW would be described. According to the mentioned hypothesis, the dimensions and intensity of those smaller streams can actually represent the basis for the prognostic modeling of not only tornadoes, but all kinds of cyclonic movements of air masses.

Similarly, studying storms in Britain, Wheeler (2001) had relied on general aspects of the procedure used by Corbyn. Those aspects are based on variations in the Sun behavior, its magnetic field, coronal eruptions and fluctuating character of the SW. Therefore, this methodology does not has almost anything in common with the majority of forecast models currently in use. The result was that 4 out of 5 strong storms have been correctly predicted in the period October 1995 – September 1997. The fifth one had a mistake of 48 hours, which can be considered as marginal (viewed from the aspect of developing methods), simply because the forecast had been done months earlier. As far as we know, the above mentioned Corbyn has not published his methods, because they were used for commercial purposes (Radovanović *et al.*, 2005).

Now, let us outline our final conclusions:

1. The formation of violent cyclonic motions and their development in the Earth's atmosphere is yet not well known and one should take into

account different conditions in the Earth's atmosphere to consider it. However, one should have in mind that the solar energy is significantly important for cyclonic motions. As we mentioned above, the most of the hurricanes appeared on geographical longitudes where sunlight reaches the Earth's surface the most effectively (longitude between 15 and 35 degrees, as described in section 2).

2. The solar wind has influence on the magnetosphere, and it seems that it may have a significant influence on the cyclonic motion in the Earth's atmosphere (see sections 2 and 3).
3. There are several violent cyclonic motions (tornadoes and hurricanes) which seem to be related to the structures on the Sun which affect the solar wind. Moreover, hurricanes Katrina, Rita and Wilma observed in 2005 appeared in the interval of time that corresponds to the solar rotation period. This may be a very strong indication that some of the violent cyclonic motions is closely related to the solar activity.
4. To understand the violent cyclonic motions (as e.g. hurricanes) one should take into account the astronomical (astrophysical) effects. Although it is not yet clear that these effects are crucial for forming the violent cyclonic motions, the investigation about possible perturbations which can occur due to the penetration of a bulk of high-energetic charged particles from the solar wind into layers of the atmosphere where cyclonic motions are starting, would be of great help to understand the phenomena. Therefore, it would be interesting to perform magneto-hydrodynamic simulations of such event.

ACKNOWLEDGMENT

The results are a part of projects III47007 and 176001 funded by the Ministry of Education and Science of the Republic of Serbia.

REFERENCES

Abdu M.A., Batista I.S., Carrasco A.J., Brum C.G.M. (2005) South Atlantic magnetic anomaly ionization: A review and a new focus on electrodynamic effects in the equatorial ionosphere, *Journal of Atmospheric and Solar-Terrestrial Physics*, 67: 1643–1657.

Arnold, N. F.; Robinson, Terry R (2001) Solar magnetic flux influences on the dynamics of the winter middle atmosphere, *Geophysical Research Letters*, 28(12): 2381-2384.

Arnold, N. F.; Robinson, Terry R (2001) Solar magnetic flux influences on the dynamics of the winter middle atmosphere, *Geophysical Research Letters*, 28(12): 2381-2384.

Antiochos, S. K.; DeVore, C. R.; Klimchuk, J. A. (1999) A Model for Solar Coronal Mass Ejections, *The Astrophysical Journal*, 510(1): 485-493.

Aschwanden, M. J. (2002) Particle acceleration and kinematics in solar flares - A Synthesis of Recent Observations and Theoretical Concepts (Invited Review), *Space Science Reviews*, 101(1): 1-227.

Acra, A., Jurdi, M., Muállem, H., Karahagopian Y, and Raffoul, Z. (1990) Water Disinfection by Solar Radiation – Assessment and Application, published by International Development Centre Canada.

Balachandran, Nambath K.; Rind, David; Lonergan, Patrick; Shindell, Drew T. (1999) Effects of solar cycle variability on the lower stratosphere and the troposphere, *Journal of Geophysical Research*, 104(D22): 27321-27340.

Barrett S. B, Leslie M. L. (2009) Links between Tropical Cyclone Activity and Madden–Julian Oscillation Phase in the North Atlantic and Northeast Pacific Basins, *Monthly Weather Review*, 137: 727-744.

Bochnícek, J.; Hejda, P.; Bucha, V.; Pýcha, J. (1999) Possible geomagnetic activity effects on weather, *Annales Geophysicae*, 17(7): 925-932.

Bossolasco, M. and Elena, A. (1972) Solar cycle control of the ionospheric E-region. *Gerlands Beitr. Geophys.*, 81: 403 – 406.

Bucha, V. and Bucha, V. (1998) Geomagnetic forcing of changes in climate and in the atmospheric circulation, *Journal of Atmospheric and Solar-Terrestrial Physics*, 60(2): 145-169.

Bucha, V. (1983) Direct relations between solar activity and atmospheric circulation, its effect on changes of weather and climate. *Stud. Geophys. Geod.*, 27: 19 – 45.

Carpenter, T. H.; Holle, R. L.; Fernandez-Partagas, J. J. (1972) Observed Relationship Between Lunar Tidal Cycles and Formation of Hurricanes and Tropical Storms, *Monthly Weather Review*. 100: 451-460.

Chapman S. and Ferraro V.C.A (1931) A new theory of magnetic storms. I. The initial phase, *Terr. Mag. and Atmosph. Elec*, 36: 77-93.

Cliver, E. W.; Boriakoff, V.; Feynman, J. (1998) Solar variability and climate change: Geomagnetic aa index and global surface temperature, *Geophysical Research Letters*, 25(7): 1035-1038.

Cohen, O.; Sokolov, I. V.; Roussev, I. I.; Arge, C. N.; Manchester, W. B.; Gombosi, T. I.; Frazin, R. A.; Park, H.; Butala, M. D.; Kamalabadi, F.; Velli, M. (2007) A Semiempirical Magnetohydrodynamical Model of the Solar Wind, *The Astrophysical Journal*, 654(2): L163-L166.

Christoforou P, Hameed S. (1997) Solar cycle and the Pacific centers of actions. *Geophysical Research Letters*, 24(3): 293-296.

D'Amicis, R.; Bruno, R.; Pallocchia, G.; Bavassano, B.; Telloni, D.; Carbone, V.; Balogh, A. (2010) Radial Evolution of Solar Wind Turbulence during Earth and Ulysses Alignment of 2007 August, *The Astrophysical Journal*, 717: 474-480.

De Jager C.; Duhau S. (2011) *The variable solar dynamo and the forecast of solar activity; Influence on terrestrial surface temperature, In Global Worming in the 21st Century*, Editor: Juliann M. Cossia, Nova Science Publisher, 77-106.

Egorova V. L, Vovk Ya V, Troshichev A. O. (2000) Influence of variations of the cosmic rays on atmospheric pressure and temperature in the Southern geomagnetic pole region, *Journal of Atmospheric and Solar-Terrestrial Physics*, 62(11): 955-966.

Elsner B. J, Kavlakov P. S. (2001) Hurricane intensity changes associated with geomagnetic variation, *Atmospheric Science Letters*, 2: 86-93.

Frank M. W, Young S. G. (2007) The Interannual Variability of Tropical Cyclones, *Monthly Weather Review*, 135: 3587-3598.

Gabis I. P, Troshichev O. A. (2000) Influence of short-term changes in solar activity on baric field perturbations in the stratosphere and troposphere, *Journal of Atmospheric and Solar-Terrestrial Physics*, 62: 725-735.

Giaiotti B. D, Giovannoni M, Pucillo A, Stel F. (2007) The climatology of tornadoes and waterspouts in Italy, *Atmospheric Research*, 83: 534–541.

Ghayoor, H. A.(1986) Temporal fluctuations of daily rainburst and the cycle of the lunar year, *International Journal of Climatology*, 6(1): 83-95.

Giorgieva K, Kirov B, Tonev P, Guineva V, Atanasov D. (2007) Long-term variations in the correlation between NAO and solar activity: The importance of north–south solar activity asymmetry for atmospheric circulation, *Advances in Space Research*, 40: 1152–1166.

Gomes J. F. P, Radovanovic M. (2008) Solar activity as a possible cause of large forest fires — a case study: Analysis of the Portuguese forest fires, *Science of the total environment*, 394(1): 197–205.

Gomes J. F. P, Radovanovic M, Ducic V, Milenkovic M, Stevancevic M. (2009) *Wildfire in Deliblatska Pescara (Serbia) - Case Analysis on July 24th 2007*. In Book: Handbook on Solar Wind: Effects, Dynamics and

Interactions. ISBN: 978-1-60692-572-0, Nova Science Publishers, New York.

Hocke K. (2009) QBO in solar wind speed and its relation to ENSO, *Journal of Atmospheric and Solar-Terrestrial Physics*, 71: 216–220.

Hodges, R. E. and Elsner, J. B (2011) Evidence linking solar variability with US hurricanes, *International Journal of Climatology*, 31(13): 1897-1907.

Jacobs, C., Poedts, S. (2011) Models for coronal mass ejections, *Journal of Atmospheric and Solar-Terrestrial Physics*, 73(10): 1148-1155.

Johnson, F.S., Mo, T., Green, A.E. (1976) Averaged latitudinal UV radiation at the earth's surface, *Photochemistry and Photobiology*, 23: 179-188.

Knabb D. R, Rhome R. J, Brown P. D. (2006) *Tropical Cyclone Report, Hurricane Katrina*, 23-30 August 2005. National Hurricane Center (http://www.nhc.noaa.gov/pdf/TCR-AL122005_Katrina.pdf).

Knabb D. R, Brown P. D, Rhome R. J. (2006) *Tropical Cyclone Report, Hurricane Rita*, 18-26 September 2005. National Hurricane Center (http://www.nhc.noaa.gov/pdf/TCR-AL182005_Rita.pdf).

Kniveton D. R, Tinsley B. A, Burns, G. B, Bering E. A, Troshichev, O. A. (2008) Variations in global cloud cover and the fair weather vertical electric field, *Journal of Atmospheric and Solar-Terrestrial Physics*, 70: 1633–1642.

Kovačević, A (2005) *Astronomy and Astrophysics*, 430: 319-325.

Kovačević, A.; Kuzmanoski, M (2007) A New Determination of the Mass of (1) Ceres, v.100, *Earth, Moon, and Planets*, 100(1-2): 117-123.

Kovačević, A (2011) *Monthly Notices of the Royal Astronomical Society*, 419(3): 2725-2736.

Legrand, J. P. and Simon, P.A. (1989) Solar cycle and geomagnetic activity: a review for geophysicists. Part I. The contributions to geomagnetic activity of shock waves and of the solar wind, *Annales Geophysicae*, 7(6): 565–578.

Leitão, P. (2003) Tornadoes in Portugal. *Atmospheric Research*, 67– 68: 381– 390.

Labitzke, K. and H. van Loon (1988) Association between the 11-year solar cycle, the QBO and the atmosphere. Part 1: The troposphere and stratosphere in the Northern Hemisphere in winter, *J. Atmos. Terr. Phys.*, 50: 197-206.

Li J. (2011) Active longitudes revealed by large-scale and long-lived coronal streamers, *The Astrophysical Journal*, 735: 130.

Lilensten J, Belehaki A. (2009) Developing the scientific basis for monitoring, modelling and predicting space weather, *Acta Geophys.*, 57:1–14.

Lilensten J, Bornarel J. (2006) *Space Weather, Environment and Societies*, Springer Verlag, Dordrecht.

Lockwood M, Stamper R, Wild N. M. (1999) A doubling of the Sun's coronal magnetic field during the past 100 years, *Nature*, 399: 437-439.

Love G. B. (2006) *Statement on Tropical Cyclones and Climate Change*. Prepared by the WMO/CAS Tropical Meteorology Research Program, Steering Committee for Project TC-2: Scientific Assessment of Climate Change Effects on Tropical Cyclones. Submitted to CAS-XIV under Agenda Item 7.3.

Lu, H., Jarvis, M. J., Graf, H.-F., Young, P. C.; Horne, R. B. (2007) Atmospheric temperature responses to solar irradiance and geomagnetic activity, *Journal of Geophysical Research*, 112(D11): CiteID D11109.

Makarov, V.I, Makarova. V.V., (1999) *Polar activity of the sun in the new global solar cycle 23*, Proc. 9th European Meeting on Solar Physics, ESA SP448, ed. A.P. Willson, 121-124.

Marhavilas P. K, Sarris E. T, Anagnostopoulos G. C. (2004) Elaboration and analysis of Ulysses' observations, in the vicinity of a magnetohydrodynamic shock, *NewsLetter and Information Service of the EGU*, 08: 30 June 2004.

Markowski M. P, Richardson P. Y. (2009) Tornadogenesis: Our current understanding, forecasting considerations, and questions to guide future research, *Atmospheric Research*, 93: 3–10.

Milovanović B, Radovanović M. (2009) The Connection between the Solar Activity and Circulation of the Atmosphere in the period 1891-2004 (in Serbian), *Journal of the Geographical institute "Jovan Cvijic" SASA*, 59/1: 35-48.

Mukherjee S, Radovanović M. (2011) Influence of the Sun in the Genesis of Tornadoes, *The IUP Journal of Earth Sciences*, 5(1): 7-21.

Nikolić J, Radovanović M, Milijašević D. (2010) An astrophysical analysis of weather based on the solar wind parameters, *Nuclear Technology and Radiation Protection*, 25(3): 171-178.

Oieroset, M.; Phan, T. D.; Angelopoulos,V.; Eastwood J. P.; McFadden J. P; Larson, D.; Carlson, C.W.; Glassmeier,K.H.; Fujimoto, M.; Raeder, J. (2008) THEMIS multispacecraft observations of magnetosheath plasma penetration deep into the dayside low-latitude magnetosphere for northward and strong By IMF, *Geophysical Research Letters*, 35: L17S11.

Oieroset, M., Phan, T. D.; Fairfield, D. H.; Raeder, J.; Gosling, J. T.; Drake, J. F.; Lin, R. P., (2008) The existence and properties of the distant magnetotail during 32 hours of strongly northward interplanetary magnetic field, *Journal of Geophysical Research*, 113(A4): A04206.

Pudovkin, M.I. and Veretenenko S.V. (1995) Cloudiness decreases associated with Forbush-decreases of galactic cosmic rays, *J. Atmos. Terr. Phys.*, 57: 1349-1355.

Palamara R. D, Bryant A. E. (2004) Geomagnetic activity forcing of the Northern Annular Mode via the stratosphere, *Annales Geophysicae*, 22: 725-731.

Pasch J. R, Blake S. E, Cobb III D. H, Roberts P. D. (2006) *Tropical Cyclone Report, Hurricane Wilma* 15-25 October 2005. National Hurricane Center, (http://www.nhc.noaa.gov/pdf/TCR-AL252005_Wilma.pdf).

Perez-Pereza J, Velasco V, Kavlakov S, Gallegos-Cruza A, Azpra-Romero E, Delgado-Delgado O, Villicana-Cruz F. (2008) *On the trend of Atlantic Hurricane with Cosmic Rays*. Proceedings of the 30th International Cosmic Ray Conference, Mexico City, 1(SH): 785–788.

Radovanović M, Stevančević M, Štrbac D. (2003) A contribution to the study of the influence of the energy of Solar wind upon the atmospheric processes, *Journal of the Geographical institute "Jovan Cvijic" SASA*, 52: 1-18.

Radovanović M, Lukić V, Todorović N. (2005) Helicentric electromagnetic long-term weather forecast and its applicable significance, *Journal of the Geographical institute "Jovan Cvijic" SASA*, 54: 5-18.

Radovanović M, Gomes J.F.P. (2009) *Solar Activity and Forest Fires*. Nova Science Publishers, New York, ISBN: 978-1-60741-002-7.

Radovanović M, (2010) Forest fires in Europe from July 22-25, 2009. *Arch. Biol. Sci.*, 62(2): 419-424.

Retinò, A.; Sundkvist, D.; Vaivads, A.; Mozer, F.; André, M.; Owen, C. J. (2007) In situ evidence of magnetic reconnection in turbulent plasma, *Nature Physics*, 3(4): 236-238.

Schielicke L, Névir P. (2009) On the theory of intensity distributions of tornadoes and other low pressure systems, *Atmospheric Research*, 93: 11–20.

Singh, A. K., Siingh, D., Singh, R. P. (2010) Space Weather: Physics, Effects and Predictability, *Surveys in Geophysics*, 31(6): 581-638.

Stevančević M, Radovanović M, Štrbac D. (2006) *Solar Wind and the Magnetospheric Door as Factor of Atmospheric Processes*. Second International Conference "Global Changes and New Challenges of 21st Century", 22-23 April 2005. Sofia, Bulgaria, 88-94.

Stevančević M. (2009) Spatial distribution of the fields with low and high atmospheric pressure as the way to the research of the geomagnetic portals (in Serbian), *Belgrade School of meteorology*, 2: 13-29.

Suparta W, Rashid Z. A. A, Ali M. A. M, Yatim B, Fraser G. J. (2008) Observations of Antarctic precipitable water vapour and its response to the solar activity based on GPS sensing, *Journal of Atmospheric and Solar-Terrestrial Physics*, 70: 1419–1447.

Svensmark H. and E. Friis-Christensen (1997) Variations of cosmic ray flux and global cloud coverage - a missing link in solar-climate relationships, *J. Atmos. Solar Terr. Phys.*, 59: 1225-1232.

Tinsley, B.A., and R.A. Heelis (1993) Correlations of atmospheric dynamics with solar activity: evidence for a connection via solar wind, atmospheric electricity, and cloud physics, *J. Geophys. Res.*, 98: 10_375-10_384.

Troshichev O. A, Janzhura A. (2004) Temperature alterations on the Antarctic ice sheet initiated by the disturbed solar wind, *Journal of Atmospheric and Solar-Terrestrial Physics*, 66: 1159–1172.

Troshichev O, Egorova L, Janzhura A, Vovk V. (2005) Influence of the disturbed solar wind on atmospheric processes in Antarctica and El Nino-Southern Oscillation (ENSO), *Mem. Soc. Astron. Ital.*, 76: 890–898.

Tyrrell J. (2007) Winter tornadoes in Ireland: The case of the Athlone tornado of 12 January 2004, *Atmospheric research*, 83: 242-253.

Vermette S. (2007) Storms of tropical origin: a climatology for New York State, USA (1851–2005), *Natural Hazards*, 42: 91–103.

Veretenenko S, Thejll P. (2004) Efects of energetic solar proton events on the cyclone development in the North Atlantic, *Journal of Atmospheric and Solar-Terrestrial Physics*, 66: 393–405.

Veretenenko S. V, Dergachev V. A, Dmitrev P. B. (2005) Long-term variations of the surface pressure in the North Atlantic and possible association with solar activity and galactic cosmic rays, *Advances in Space Research*, 35: 484-490.

Visvanathan, T. R (1966) Formation of Depressions in the Indian Seas and Lunar Phase, *Nature*, 210(5034): 406-407.

von Storch H, Zwiers F. W. (1999) *Statistical analysis in climate research*. Cambridge University Press, Cambridge, UK.

Yaukey, P. (2009) *Hurricane Rapid Growth Events and the Lunar Synodic Cycle*. Paper presented at Hurricanes III Climate Dynamics and Biotic Response. Association of American Geographers 2009, Annual General Meeting. March 2009. Las Vegas, Nevada.

Watermann, J.; Wintoft, P.; Sanahuja, B.; Saiz, E.; Poedts, S.; Palmroth, M.; Milillo, A.; Metallinou, F.-A.; Jacobs, C.; Ganushkina, N. Y et al. (2009) Models of Solar Wind Structures and Their Interaction with the Earth's Space Environment, *Space Science Reviews*, 147(3-4): 233-270.

Wheeler D. (2001) A verification of U. K. gale forecasts by the 'solar weather technique': October 1995–September 1997, *Journal of Atmospheric and Solar-Terrestrial Physics*, 63(1): 29-34.

Zweibel, E. G. and Yamada, M. (2009) Magnetic Reconnection in Astrophysical and Laboratory Plasmas, *Annual Review of Astronomy and Astrophysics*, 47(1): 291-332.

In: Solar Wind ISBN 978-1-62081-979-1
Editors: C. D. E. Borrega, A. F. B. Cruz ©2012 Nova Science Publishers, Inc.

Chapter 2

PLANETARY BEAT, SOLAR WIND AND TERRESTRIAL CLIMATE

Nils-Axel Mörner[*]
Paleogeophysics and Geodynamics, Stockholm, Sweden

ABSTRACT

The Sun's emission of luminosity and Solar Wind is constantly changing. This solar variability is driven by a planetary beat because the Solar-Planetary system behaves like a multi-body system constantly adjusting their motions around a common centre of gravity. The solar influence on terrestrial climate and environments seems primarily to go via the interaction of the Solar Wind with the Earth's magnetosphere. This generates changes in shielding capacity, geomagnetic activity, atmospheric pressure, external gravity and Earth's rate of rotation. The correlation between changes in solar activity and Earth's rate of rotation is indicative of a forcing function via changes in Solar Wind emission (not luminosity). Changes in rotation (LOD) affects not only the atmospheric circulation but also the ocean circulation. Changes in ocean circulation lead to the redistribution of oceanic water masses (recorded by sea level changes) and water-stored heat (recorded by paleoclimate). Changes in ocean circulation have strong effects on terrestrial climate. During Grand Solar Minima (the Spörer, Maunder and Dalton Minima) the Earth's experienced a speeding-up in the rate of rotation as evidenced by significant reorganisations of the currents in the North Atlantic. This

[*] E-mail: morner@pog.nu.

lead to the establishment of periods of cold climatic conditions, known as Little Ice Ages. The next Grand Solar Minimum is due at about 2030-2040. We must assume that this will be a period of resumed cold climatic conditions, may be even Little Ice Age conditions.

Keywords: Solar Wind, Planetary beat, Solar-terrestrial interaction, Earth's rotation, ocean circulation, Solar Maxima and Minima

INTRODUCTION

The Sun's activity is constantly changing. These changes follow cyclic pattern of fairly strict frequency. The 11 years sunspot cycle may be the most well known cycle. There are both shorter and longer cycles, however. The origin of the cyclic behaviour in the Sun's variability is open for different interpretations.

The constantly adjusting motions of the planets around a common centre of mass generates a gravitational beat within the planetary system, which induces cyclic pulses on the Sun that are of similar frequency as those recorded in the Sun's activity variations (e.g. Landscheidt, 1976; Mörth and Schlamminger, 1979; Fairbridge, 1984; Mörner, 1984; Scafetta, 2010; Fix, 2011).

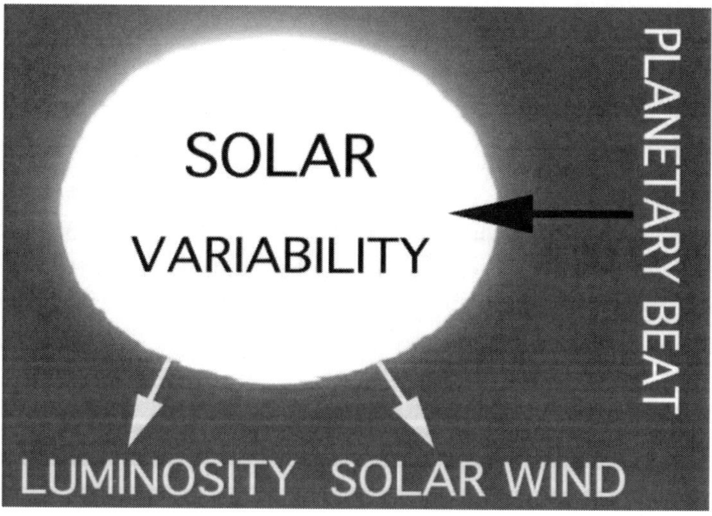

Figure 1. Relations, here proposed, among planetary beat, solar variability and emission of luminosity and Solar Wind.

Scafetta (2010) talks about "astronomical oscillations" which is the same as my "planetary beat" (Mörner, 1984). He established an excellent correlation between the 60 years terrestrial LOD cycle and the 60 years cycle of changes in the orbital speed of the Sun around the centre of mass of the solar system (Scafetta, 2010, Figure 14), in line with the causal connections here proposed (Figure 1).

In conclusion, a planetary beat (astronomical oscillation) as ultimate driver of the variability in Solar activity seems much more reasonable than "a chronometer hidden in the Sun" (Dicke, 1978).

This is illustrated in Figure 1, where the changes in solar activity are held to be controlled by the planetary beat and where the out-going variables are the luminosity (light and energy) and the Solar Wind (solar-magnetic plasma and particles).

Planetary Beat

Instead of fixed, "Keplerian", motions of the planets around the Sun, the solar-planetary system behaves like a multi-body system in its constantly adjusting motions around the common centre of gravity. This implies that all the planetary bodies involved, including the Sun, are affected by irregular gravitational and angular momentum forces; "a planetary beat" (Mörner, 1984) or "astronomical oscillation" (Scafetta, 2010). Most of the forces (99%) are generated by the big planets Jupiter (61.2%), Saturn (24.6%), Neptune (8.0%) and Uranus (5.5%). Their beat on the Sun is proposed to generate the changes of solar activity recorder in the solar cycles (Mörth and Schlamminger, 1979) and illustrated in Figures 1 and 2.

The planetary beat also affects the Earth and the Earth-Moon system (Mörner, 1984) as illustrated in Figure 2. A very significant influence on Planet Earth seems to come from the interaction of the Solar Wind with the Earth's magnetosphere.

Solar Wind

The Sun emits luminosity and Solar Wind. Whilst the variations in luminosity over a sunspot cycle is quite low (in the order of 0.2%) the changes in Solar Wind emission are quite strong over a sunspot cycle: so for example, the velocity varying between ~400 km s^{-1} at slow solar wind and ~750 km s^{-1} at fast solar wind. The Solar Wind is here claimed to have the predominant controlling effect not only on the space environment and weather of Planet Earth but also on the solar impact on Earth's climate and global environments (Mörner, 2010, 2011) as illustrated in Figure 2.

The interaction of the Solar Wind with the Earth's magnetosphere (Figure 2) affects (1) the Earth's shielding capacity towards cosmic rays and by that probably the cloud formation, (2) the geomagnetic intensity and by that the electric circuit in the ionosphere, (3) the atmospheric pressure distribution and by that the wind systems, (4) the gravity distribution, and (5) the Earth's rate of rotation and by that the atmospheric and ocean circulation systems. From this lead three ways to changes in the terrestrial climate and environments; viz: clouds, winds and ocean circulation.

1. The Shielding Capacity

It is a well-known fact that the shielding capacity of the Earth's magnetosphere against cosmic rays varies with the Solar Wind changes. These variations are monitored by the production of ^{14}C in the atmosphere (e.g. Stuiver and Quay, 1980) and by the in-fall of ^{10}Be at the Earth's surface (e.g. Bard et al., 2000).

According to Svensmark (e.g. 1998, 2007) the atmospheric flux of cosmic rays is likely also to control the formation of clouds; generating more clouds during stronger flux and less clouds during weaker flux.

2. Geomagnetics

The Solar Wind has a direct effect on the Earth's magnetosphere. Hence it is not surprising that there is a correlation among solar activity and geomagnetic activity (Cliver et al., 1998). Stronger and weaker geomagnetic activity makes the Arctic circulation to switch between zonal and meridional circulation according to Bucha (e.g. 1984), today known as the Arctic Oscillation or AO (Thompson and Wallace, 1998).

It also affects the electric circuit in the ionosphere (Boberg and Lundstedt, 2002), which is another line of affecting the global wind pattern, and by that climate and environment on the Earth (Figure 2).

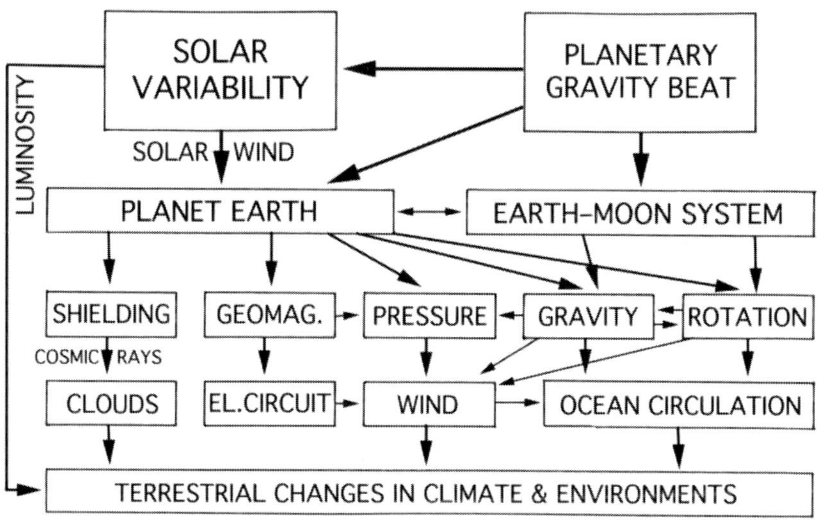

Figure 2. The interaction between planetary beat and solar variability, and changes in climate and environments on Planet Earth.

The Earth's geomagnetic field is controlled partly by the Earth's own internal geomagnetic field, and partly by the external contribution from the Heliomagnetic field via the Solar Wind (as illustrated with opposed arrows in Figure 3).

De Santis et al. (2012) have recently suggested that there is a causal linkage between the spatial growth since AD 1600 of the geomagnetic field low anomaly over the South Atlantic (SAA) and global changes in sea level. The sea level data compared with (Jevrejeva et al., 2008) do, however, not represent a firm solution of the actual changes in global sea level (better summarized in Figure 3 of Mörner, 2004). At the same time any dislocation at the core/mantle boundary (Figure 3) will lead to deformations of the sea surface topography as proposed by Mörner (1980).

3. Pressure

The atmospheric pressure distribution is affected by the Solar Wind interaction with the magnetosphere and by the planetary gravitational beat on Planet Earth and the Earth-Moon system (Figure 2). Any redistribution in the pressure pattern leads to readjustments in the wind pattern.

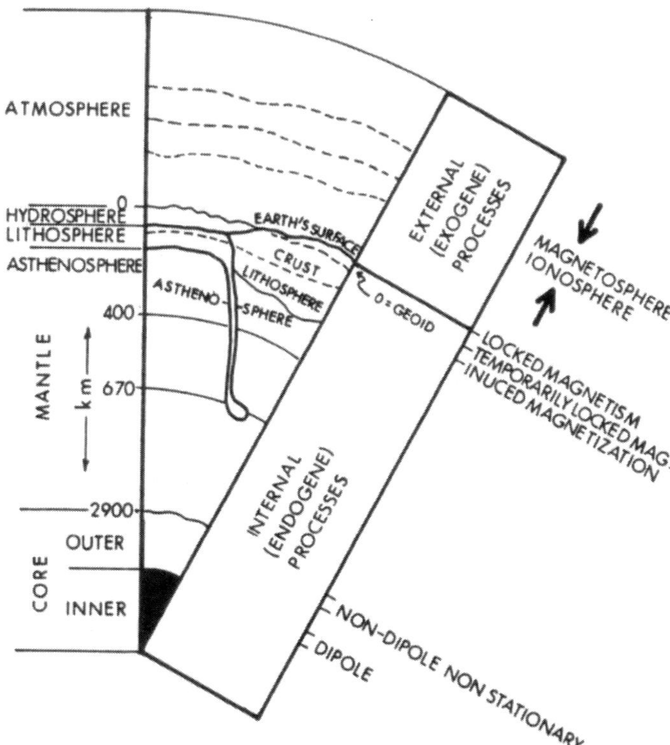

From Mörner, 1989a.

Figure 3. A segment of the Earth with different boundaries and layers, which may be affected by differential rotation.

4. Gravity

The planetary beat on Planet Earth and the Earth-Moon system generates gravity pulses, which are picked up in the configuration of the equipotential surfaces and in the tidal forces, by that having the power of affecting the atmospheric and oceanic circulation systems (Figure 2).

5. Rotation

The Earth's rate of rotation is constantly changing due to a number of different reasons; viz. long-term tidal friction (e.g. Marsden and Cameron,

1966), core/mantle interaction and geomagnetics (e.g. Braginsky, 1982; Rochester, 1984; Roberts et al., 2007), plate motions (Mörner, 1989a), ocean circulation (Mörner, 1984, 1988, 1990, 1992), atmospheric circulation (e.g. Hide and Dickey, 1990; Le Mouël et al., 2010), major earthquakes and volcanic eruptions, etc.

Figure 3 gives a segment of the Earth from core to atmosphere with different layers and sub-layers marked, along which differential rotation may take place (Mörner, 1996a, Table 1).

Several authors have noted a correlation between sunspot activity and Earth's rate of rotation (e.g. Kalinin and Kiselev, 1976; Golovkov, 1983; Mazzarella and Palumbo, 1988; Gu, 1998; Rosen and Salstein, 2000; Kirov et al., 2002; Abarca del Rio et al., 2003; Mazzarella, 2007, 2008; Mörner, 2010, Le Mouël et al., 2010; Scafetta, 2010, 2012).

These correlations may seem incomprehensible from the view of changes in solar luminosity emission. It terms of corresponding changes in the Solar Wind emission and their interaction with the Earth's magnetosphere, the correlations becomes quite reasonable, however (Mörner, 1996a).

The Earth's rate of rotation is usually measured as changes in the length of the day (LOD), where a speeding up or acceleration lead to a decrease in LOD and a slowing down or deceleration leads to an increase in LOD.

The correlation between solar activity and LOD includes annual, decadal, multi-decadal and centennial signals. Mörner (2010, Figure 2) noted a correlation between annual sunspot numbers and LOD values from 1831 to 1995. Several authors have recorded a correlation between the sunspot cycle and LOD; e.g. Kalinin and Kiselev, 1976; Golovkov, 1983; Mazzarella and Palumbo, 1988; Abarca del Rio et al., 2003; Le Mouël et al. (2010). Kirov et al. (2002) found a correlation between the 22-years Hail cycle and LOD. Mazzarella (2007, 2008) and Scafetta (2010) documented a close correlation between the 60-years cycle in solar activity and in LOD. On the longer-term basis, Mörner (2010, 2011) showed the Grand Solar Minima of the Spörer Minimum, the Maunder Minimum and the Dalton Minimum corresponded to periods of speeding-ups in the Earth's rate of rotation, whilst the Solar Maxima corresponds to slowing-downs in the Earth's rate of rotation (as further discussed below and illustrated in Figures 4-5).

The correlation between changes in solar activity and Earth's rate of rotation (Figure 2) can only be understood in terms of Solar Wind interaction with the Earth's magnetosphere (Mörner, 1996a). This is strongly supported also by the correlations established between solar activity and cosmogenic isotopes (e.g. Stuiver and Quay, 1980; Bard et al., 2000; Bond et al., 2001), solar activity and geomagnetics (e.g. Bucha, 1984; Krivova et al., 2007; Mufti and Shah, 2011), and Solar Wind variations and various terrestrial parameters (e.g. Boberg and Lundstedt, 2002).

6. Impact

The variations in solar activity lead to the emission of luminosity as well as of Solar Wind (Figure 1). Whilst the changes in luminosity are too small to have any major effects on the terrestrial systems, the variations Solar Wind, via its interaction with the Earth's magnetosphere, lead to quite drastic changes (Figure 2), which are likely to generate significant changes in Earth's climate and environments (Mörner, 1996a, 2010).

Solar-Terrestrial Interaction

There seems to be four lines (Figure 2), along which solar variability may affect the terrestrial changes in climate and environments; viz. (1) direct luminosity effects, (2) variations in cloud formation according to the Svensmark theory, (3) variations in atmospheric circulation and wind, and (4) variations in ocean circulation.

Before discussing these different lines of solar-terrestrial interaction, it must be emphasized that there has often been an incorrect attribution of recorded changes as originating from "solar irradiance" when they in fact originate from "Solar Wind".

The "solar irradiance curve" of Bard et al. (2000) should rather be labelled "a Solar Wind curve" (below Figure 5; Mörner, 2010, 2011) as it is constructed from the variations in cosmogenic nuclides controlled by the variations in shielding capacity of the Earth's geomagnetic field.

Similarly, the "reconstruction of total irradiance since 1700" by Krivova et al. (2007; cf. Scafetta, 2011, Figure 15) is based on surface magnetic flux, which is strongly controlled by the Solar Wind interaction with the magnetosphere and has nothing to do with changes in irradiance (Figure 2).

The excellent correlation for the last 150 years between changes in the length of the solar cycle and global mean temperature (Friis-Christensen and Lassen, 1991) may, theoretically, refer to solar irradiance as well as Solar Wind (Figure 2).

Eddy (1976) proposed that Little Ice Ages occurred at periods of solar minima. Indeed, there are good correlations between cooling events over northern Europe, the North Atlantic and the Arctic and the timing of the Spörer, Maunder and Dalton Grand Solar Minima. During these minima the Earth experienced speeding-up phases in the rate of rotation due to decreases in Solar Wind emission during Solar Minima (Mörner, 2010, 2011; cf. below: Figures 4-5).

1. Luminosity

Very much has been written about possible direct effect on Earth's climate from changes in the solar energy output. The variability between the sunspot cycles seems far too small, however (e.g. Willson, 1997). One way of trying to overcome this problem has been to assume hypothetical amplifying factors (e.g. Hoyt and Schatten, 1993; Lean et al., 1995; Lean and Rind, 1999).

The correlations observed between changes in solar variability and various climatic parameters seem much better understood in terms of effect via the Solar Wind emission, however (Figure 2).

This dose not mean that direct solar luminosity must have zero effects on Earth's climate, only that these effects must be small to subordinate.

2. Cloud Formation

Variations in cloud formation as a function of a variable flux of cosmic rays in the upper atmosphere (due to changes in the shielding capacity generated by the Solar Wind interaction with the magnetosphere) have been proposed by Svensmark (e.g. 1998, 2007; Svensmark and Calder, 2007). This may well be an important way of modulating Earth's climate.

3. Wind

The global wind pattern and atmospheric circulation have a strong controlling function on regional and global climate. Jelbring (1998) even talks about "wind driven climate". The annual LOD changes seem well balanced by changes in atmospheric angular momentum (e.g. Barnes et al., 1983). For longer-term LOD changes, the oceanic circulation and core/mantle coupling must also be considered. The speed and geographic pattern of the jet streams are strongly linked to changes in LOD. The alternations between zonal and meridional circulation in the Arctic are linked to changes in geomagnetic activity (Bucha, 1983, 1984) where the solar forcing must be driven by the Solar Wind variability (Figure 2). These changes are today known as the Arctic Oscillation or AO (Thompson and Wallace, 1998).

Boberg and Lundstedt (2002) proposed that the North Atlantic Oscillation was a function changes in the electric circuit induced by Solar Wind interaction with the magnetosphere (Figure 2).

Le Mouël et al. (2010) showed that there is a good correlation between the sunspot cycles, the flux of cosmic rays in the upper atmosphere and LOD. This is indicative of a solar-terrestrial linkage via changes in the Solar Wind (cf. above and Figure 2). They assumed that the changes in LOD were a function only of changes in atmospheric circulation. Similarly, Mazzarella (2007, 2008) and Scafetta (2010) discussed the 60-years cycle in terms of atmospheric circulation (cf. below).

Whilst the feedback coupling between changes in atmospheric circulation and LOD usually seems to be balances on the inter- to intra-annual basis (e.g. Barnes et al., 1983), the feedback coupling between changes in ocean circulation and LOD usually takes decades to become balanced (Mörner, 1984, 1995a).

4. Ocean Circulation

Ocean circulation is usually discussed in terms of thermohaline circulation. The global circulation systems of oceanic surface currents must, however, also be strongly dependent on the Earth's rate of rotation (Mörner, 1984). The equatorial currents are lagging-behind the rotation of the solid Earth, and the Gulf Stream and Kuroshio Current bring hot equatorial water from low latitudes to high latitudes in the North Atlantic and Pacific, respectively, which has to have a strong feedback coupling to the rate of rotation (Mörner, 1984, 1988, 1990).

The El Niño/ENSO events include a transfer of angular momentum from the solid Earth to the hydrosphere and back again, allowing the equatorial current in the Pacific to reverse its direction and transport hot water eastwards, blocking the cold Humboldt Current and rising sea level some 30 cm along the American west coasts (Mörner, 1989b, 1996a, Figure 10, 2012, Figure 4).

In the North Atlantic, Mörner (1984, 1995a) recorded 16 pulses in marine biota, sea level changes and regional temperature, which he interpreted in terms of an interchange of angular momentum between the hydrosphere and the solid earth in a feedback coupling generating pulses in the Gulf Stream beat. Similar pulses were recorded in the Kuroshio Current, and in a back-and-forth wave of equatorial water masses in the Pacific–Indian Ocean (Mörner, 1993a, 1995b).

Whilst the global sea level changes prior to ~6000 BP were dominated by the glacial eustatic rise in sea level, they were during the last 6000 years dominated by these irregular relocations of ocean water masses, which obviously must be linked (in a feedback coupling) to changes in Earth's rate of rotation, ultimately driven by solar–planetary changes (as illustrated in Figures 1 and 2). During the last glaciation maximum (LGM) some 20,000 years ago when sea level was in the order of 130 m lower, Earth's rate of rotation was considerably faster, giving rise to a significantly increase in the lagging-behind of the equatorial currents (Mörner, 1995a, 1996a). Even during the deglacial phase there were large-scale changes in ocean circulation and rate of rotation affecting the course and intensity of the Gulf Stream and Kuroshio Current (Mörner, 1993b, 1995b, 1996a, 1996b). This also offers a way of understanding long-distance correlations with changes in the Indian Summer monsoon as recorded in cores in the Indian Ocean (Kessarkar et al., 2011).

The 60-years cycle recorded in solar activity and LOD (Mazzarella, 2007, 2008; Scafetta, 2010) must certainly also have affected the oceanic circulation, as it is found in different marine environments scattered widely over the globe (e.g. Black et al., 1999; Patterson et al., 2004; Klyashtorin et al., 2009; Mörner, 2012). The Pacific Decadal Oscillation (PDO) has a 60-years periodicity. Like the ENSO events (cf. above), it seems also to include an oceanic component of interchange of angular momentum.

The data from the Arctic (Klyashtorin et al., 2009) seem especially relevant. The stocks of herring and cod in the Barents Sea fluctuate with a 60-years periodicity, which correlates with the changes in Arctic temperature, in ocean water temperature and in ice cover conditions in the Barents Sea; all exhibiting an ~60-years periodicity. Klyashtorin et al. (2009) noted that the changes in ice cover reflect the "delivery of warm Atlantic water to the region" and that "the main source of heat delivered to the Arctic basin is warm water inflow from the North Atlantic Stream". This implies that the Gulf Stream system must exhibit a beating cycle of 60 years, just as the LOD cycle and solar activity cycle, in full agreement with the proposal of Mörner (1984, 2010, 2011, and illustrated in 2012, Figure 5).

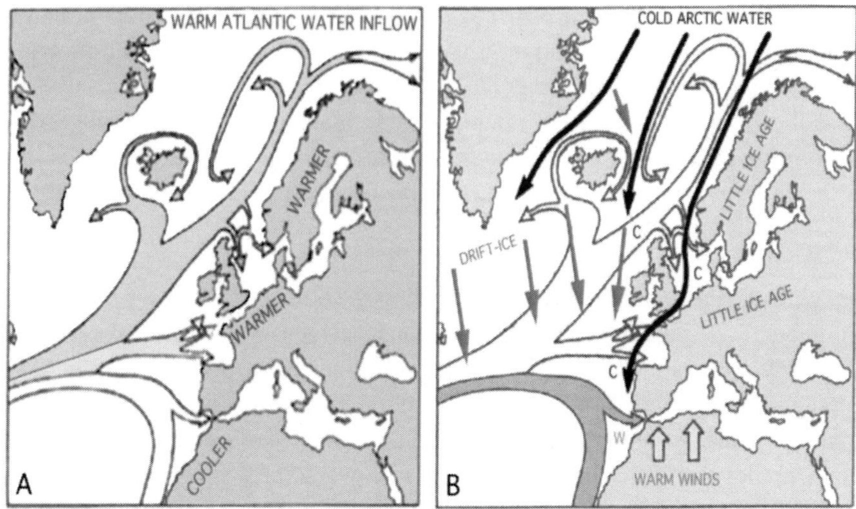

Figure 4. The Gulf Stream and its changes in distribution of water masses at Grand Solar Maxima (A) and Grand Solar Minima (B). During the Spörer, Maunder and Dalton Solar Minima (B) Little Ice Age conditions prevailed in northwest Europe. The next Grand Solar Minimum is due at around 2030-2040 (Figure 5).

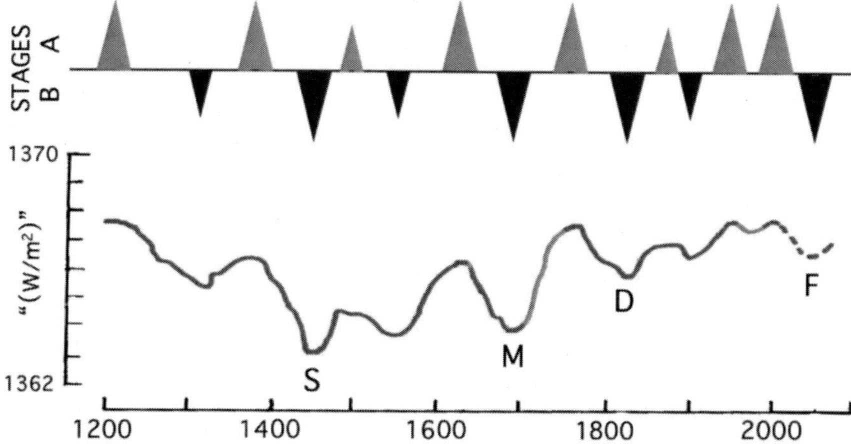

Figure 5. Solar variability (below) and Atlantic stages A and B periods (above). The "solar irradiance" curve (below) of Bard et al. (2000) records the changes in Solar Wind, not irradiance, and must hence be relabelled "a curve of the changes in Solar Wind activity" (therefore their vertical scale of irradiance is put in brackets and quotation marks). The upper graph gives the changes between stage A and stage B conditions in the North Atlantic region as given in Figure 4.

The Gulf Stream system seems extremely sensitive to changes in the rate of rotation. Any change in rotation may be compensated by a change in water mass transport within the Gulf Stream system, and, vice versa, any change in the water transport (volume as well as direction) should affect the Earth's rate of rotation (Mörner, 1984). The transport of warm equatorial water along the Gulf Stream may be predominantly directed along its northern branches to northwestern Europe and the Arctic basin (stage A in Figure 4), or along the southern branch to southwest Europe and northwest Africa (stage B in Figure 4) as shown by Mörner (1996a, 2010). Paleoclimatic time series with an annual resolution made it possible to record the temporal and spatial changes in climate for the last millennium over the east Atlantic to west European region (Mörner, 2010). During the Spörer, Maunder and Dalton Solar Minima Arctic water penetrated all the way down to mid-Portugal and the Gulf Stream transport was concentrated along the southern branch (stage B in Figure 4). This lead to Little Ice Age conditions in the north and central Europe, and opposed warming conditions in Gibraltar region and Northwest Africa.

During the Solar Maxima the situation was reversed; the warm Atlantic water was transported far up into the Barents Sea region, making west Europe and the Arctic unusually warm, whilst the Gibraltar region and northwest Africa suffered cool conditions because of decreased transport along the southern branch of the Gulf Stream (stage A in Figure 4). The switches between those two modes of ocean circulation in the North Atlantic are driven by the speeding-up of the Earth's rotation at Solar Minima and the slowing-down at Solar Maxima as a function of the variability in the Solar Wind emission and its interaction with Earth's magnetosphere (as illustrated in Figure 2, and Mörner, 1996a, 2010, 2011, 2012).

Figure 5 gives the changes in Solar Wind activity during the last 800 years as established by Bard et al. (2000) and termed "solar irradiance" though it, in fact, is a Solar Wind activity curve (Mörner, 2010). The alternation between solar maxima and solar minima (Spörer, Maunder, Dalton) is clearly recorded. The corresponding changes in the Atlantic circulation with switches between stages A and B conditions (Figure 4) fit the changes in Solar Wind beat very well as show in the upper graph of Figure 5.

5. Integration of Variables

Surely Earth's climate is affected by changes in atmospheric circulation as well as by changes in ocean circulation. Variations in cloud formation in response to the Solar Wind modulation of Earth's shielding capacity and by that the flux of cosmic rays in the atmosphere may have significant climatic effects (as proposed by Svensmark, 1998, 2007). Direct effects from the changes in luminosity are considered to be small.

Conclusion

The solar influence on terrestrial climate and environment primarily goes via the interaction of variations in the Solar Wind with the Earth's magnetosphere and its effects on shielding, geomagnetism, pressure, gravity and especially rotation. Direct effects from the changes in luminosity seem to be small to subordinate.

Changes in the ocean circulation seem closely linked to changes in Earth's rate of rotation (LOD). The correlations recorded between changes in solar activity and LOD must go via the Solar Wind and its effects on the rate of rotation, and the Planetary beat (Figures 1 and 2).

The changes in ocean circulation have strong effects on the terrestrial climate, especially in relation to the changes between at the Grand Solar Minima (the Spörer, Maunder and Dalton Minima) with Little Ice Age conditions, and Grand Solar Maxima with warm climate conditions (as illustrated in Figure 4; cf. Mörner 1996a, 2010, 2011).

The next Grand Solar Minimum has been extrapolated to occur at around 2030 by Landscheidt (2003), at around 2040 by Mörner et al. (2003), at around 2030 by Klyashtorin et al. (2009), at 2030-2040 by Harrara (2010), at 2042 ±11 by Abdassamatov (2010), at 2030-2040 by Easterbrook (2011), at around 2035 by Malberg (2012) and at around 2030 by Lyubushin and Klyashtorin (2012), implying a fairly congruent picture despite quite different ways of transferring past signals into future predictions. In analogy with the past Solar Minima, one may assume that the future minimum at around 2030-2040 will also generate Little Ice Age climatic conditions (Figure 4).

Acknowledgments

The theory of an interchange of angular momentum between the solid Earth and the hydrosphere was first presented at the international symposium on Climatic Changes on a Yearly to Millennial Basis in Stockholm in 1983, and at the IAMAP/IAPSO conference in Hawaii in 1985. The impact of Solar Wind changes on Earth's climate was presented at the IUGG conference in Birmingham in 1999. The rotational changes in relation to Solar Maxima/Minima alterations and its impact for future climate was presented at the EGS/AGU/EUG conference in Nice in 2003. References are also made to the INTAS 97-31008 project on "Geomagnetism and Climate", co-ordinated by the present author from the department of Paleogeophysics and Geodynamics at Stockholm University. Figure 1 was constructed on the basis of a photography of the setting Sun in SW Sweden taken by my friend Dr. Bjarne Lembke. I am indebted to Professor Don J. Easterbrook for constructive reviewing of the paper.

REFERENCES

Abarca del Rio, R., Gambis, D., Salstein, D., Nelson, P. and Dai, A. (2003). Solar activity and earth rotation variability. *J. Geodynamics*, 36, 423-443.

Abdassamatov, H.I. (2010). The Sun dictates the climate. ppt-presentation. http:/www.heartland.org/events/2010Chicago/program.html

Bard, E., Raisbeck, G., Yiou, F. and Jouzel, J. (2000). Solar irradiance during the last 1200 years based on cosmogenic nuclides. *Tellus*, 52B, 985-992.

Barnes, R.T.H., Hide, R., White, A.A. and Wilson, C.A. (1983). Atmospheric angular momentum correlated with length of the day changes and polar motion. *Proc. Roy. Soc. London*, Ser. A, 387, 31-73.

Black, D.E., Peterson, L.C., Overpeck, J.T., Kaplan, A., Evans, M.N. and Kashgarin, M. (1999). Eight centuries of North Atlantic ocean-atmosphere variability. *Science*, 286, 1709-1713.

Boberg, F. and Lundstedt, H. (2002). Solar Wind variations related to fluctuations of the North Atlantic Oscillation. *Geophys. Res. Letters*, 29 (15), p. 13:1-4, 10.1029/ 2002GL014903.

Bond, G., Kromer, B., Beer, J., Muscheler, R., Evans, M.N., Showers, W., Hoffmann, S., Lotti-Bond, R., Hajdas, I. and Bonani, G. (2001). Persistent solar influence on North Atlantic climate during the Holocene. *Science* 294, 2130-2136.

Braginskiy, S.I. (1982). Analytical description of the secular variations of the geomagnetic field and the rate of rotation of the Earth. *Geomagnetism Aeronomy*, 22, 88-94.

Bucha, V. (1983). Direct relations between solar activity and atmospheric circulation, its effects on weather and climate. *Studia geophysica et geodetica*, 27, 19-45.

Bucha, V. (1984). Mechanism for linking solar activity to weather-scale effects, climatic changes and glaciations in the Northern Hemisphere. In: *Climatic Changes on a Yearly to Millennial Basis* (N.-A. Mörner and W. Karlén, Eds), p. 415-448. Reidel Publ. Co. (Dordrecht/Boston/Lancaster).

Cliver, E.W., Boriakoff, V. and Feynman, J. (1998). Solar variability and climate changes: geomagnetic and aa index and global surface temperature. *Geophys. Res. Lett.* 25, 1035-1038.

De Santis, A., Qamili, E., Spada, G. and Gasperini, P. (2012). Geomagnetic South Atlantic Anomaly and global sea level rise: A direct connection? *J. Atmospheric and Solar-Terrestrial Physics*, 74, 129-135.

Dicke, R.H. (1978). Is there a chronometer hidden in the Sun? *Nature*, 276, 676-680.

Easterbrook, D.J. (2011). Geologicl evidence of recurring climate cycles and their implications for the cause of global climate change – the past is the key to the future. In: *Evidenced-based Climate Science* (D.J. Easterbrook, Ed.), p. 3-51, Elsevier.

Eddy, J.A. (1976). The Maunder Minimum. *Science* 192, 1189-1202.

Fairbridge, R.W. (1984). Planetary periodicities and terrestrial climate stress. In: *Climatic Changes on a Yearly to Millennial Basis* (N.-A. Mörner and W. Karlén, Eds), p. 509-520. Reidel Publ. Co. (Dordrecht/Boston/Lancaster).

Fix, E.L. (2011). The relationship of sunspot cycles to gravitational stresses on the Sun: results of a proof-of-concept simulation. In: *Evidenced-based Climate Science* (D.J. Easterbrook, Ed.), p. 335-350, Elsevier.

Friis-Christensen, E. and Lassen. K. (1991). Length of the Solar cycle: an indication of Solar activity closely associated with climate. *Science* 254, 698-700.

Golovkov, V.P. (1983). Dynamics of the geomagnetic field and the internal structure of the Earth, In: *Magnetic field and processes in the Earth's interior* (V. Bucha, Ed.), p. 395-501. Czeck Acad. Sci., Academia, Pragúe.

Gu, Z. (1998). An interpretation of the non-tidal secular variation in the earth rotation: the interaction between the solar wind and the earth's magnetosphere. *Astrophys. Space Sci.*, 259, 427-432.

Harrara, V.M.V. (2010). The New Solar Minimum and the Mini Ice Age of the Twenty-First Century. ppt-presentation.

http:/www.heartland.org/events/2010Chicago/program.html

Hoyt, D.V. and Schatten, K.H. (1993). A discussion of plausible solar irradiance variations, 1700-1992. *J. Geophys. Res.* 98, 18,895-18,906.

Jelbring, H. (1998). Wind driven climate. Ph.D. thesis, No. 9, Paleogeophysics and Geodynamics, Stockholm University.

Jevrejeva, S., Moore, J.C., Grinsted, A. and Woodworth, P.L. (2008). Recent global sea level acceleration started over 200 years ago? *Geophys. Res. Letters*, 35, L08715, doi:10.1029/ 2008GL033611

Kalinin, Y.D. and Kiselev, V.M. (1976). Solar conditionality of changes in the length of the day, the seismicity of the Earth and the geomagnetic moment. *Geomag. Aeromag.* 16, 858-870.

Kessarkar, P.M., Roa, V.P., Naqvi, S.W.A. and Karapurkar, S.G. (2011). Variatiions in the Indian Summer Monsoon during the Bolling_Alleröd and Holocene. *7th Intern. Conf. Asia Marine Geology*, Goa, Abstracts, p. 89.

Kirov, B., Georgieva, K. and Javaraiah, J. (2002). 22-year periodicity in solar rotation, solar wind parameters and earth rotation. Proc. 10th European Solar Physics Meeting, Prague, p. 149-152.

Klyashtorin, L.B., Borisov, V. and Lyubushin, A. (2009). Cyclic changes of climate and major commercial stocks of the Barents Sea. *Marine Biology Res.*, 5, 4-17.

Krivova, N.A., Balmaceda, L. and Solanki, S.K. (2007). Reconstruction of solar total irradiance since 1700 from surface magnetic flux. *Astronomy Astrophycics*, 467, 335-346.

Landscheidt, T. (1976). Beziehungen zwischen der Sonnenaktivität und dem Massen-zentrum des Sonnensystem. *Nachr. Olders-Ges. Bremen*, 100, 2-19.

Landscheidt, T. (2003). New Little Ice Age instead of Global Warming. *Energy and Environment*, 14, 327-350.

Lean, J., Beer, J. and Bradley, R. (1995). Reconstruction of solar irradiance since 1610: implications for climate change. *Geophys. Res. Lett.* 22, 3195-3198.

Lean, J. and Rind, D. (1999). Evaluating sun-climate relationships since the Little Ice Age. *J. Atmosph. Solar-Terrest. Physics*, 61, 25-36.

Le Mouël, J.L., Blanter, E, Shnirman, M. and Cortillot, V. (2010). Solar forcing of the semi-annual variation of length-of-day. *Geophysical Research Letters*, 37, L15307, 5 pp. doi:10.1029/2010GL043185

Lyubushin, A.A. and Kluashtorin, L.B.(2012). Short term dT prediction using (60-70)-years periodicity. *Energy and Environment*, 23 (1), 74-85.

Malberg, H. (2012). *Beiträge zur Berliner Wetterkarte*, Über sprunghafte Anstiege von CO2 und globaler Temperatur. http://www.Berliner-Wetterkarte.de

Marsden, G.B. and Cameron, A.G.W. (1966). *The Earth-Moon System*. Plenum Press, New York.

Mazzarella, A. (2007). The 60-year solar modulation of global air temperature: the Earth's rotation and atmospheric circulation connection. *Theoretical and Applied Climatology*, 88, 193-199.

Mazzarella, A. (2008). Solar forcing of changes in atmospheric circulation, Earth's rate of rotation and climate. *The Open Atmospheric Science Journal*, 2, 181-184.

Mazzarella, A. and Palumbo, A. (1988). Earth's rotation and solar activity. *Geophysical Journal*, 97, 169–171.

Mörner, N.-A. (1980). Eustasy and geoid changes as a function of core/mantle changes. In: *Earth Rheology, Isostasy and Eustasy* (N.-A. Mörner, ed.), p. 535-553.

Mörner, N.-A. (1984). Planetary, solar, atmospheric, hydrospheric and endogene processes as origin of climatic changes on the Earth. In: *Climatic Changes on a Yearly to Millennial Basis* (N.-A. Mörner and W. Karlén, Eds), p. 483-507. Reidel Publ. Co. (Dordrecht/ Boston/Lancaster).

Mörner, N.-A. (1988). Terrestrial variations within given energy, mass and momentum budgets; Paleoclimate, sea level, paleomagnetism, differential rotation and geodynamics. In: *Secular Solar and Geomagnetic Variations in the last 10,000 years* (F.R. Stephenson and A.W. Wolfendale, Eds.), p. 455-478, Kluwer Acad. Press.

Mörner, N.-A. (1989a). Global Change: The lithosphere: Internal processes and Earth's dynamicity in view of Qauternary observational data. *Quaternary International*, 2, 55-61.

Mörner, N.-A. (1989b). Changes in the Earth's rate of rotation on an El Niño to century basis. In: *Geomagnetism and Paleomanetism* (F.J. Lowes et al., Eds), p. 45-53, Kluwer Acad. Publ.

Mörner, N.-A. (1990). The Earth's differential rotation: hydrospheric changes. *Geophysical Monographs*, 59, 27-32, AGU and IUGG.

Mörner, N.-A., (1992). Sea level changes and Earth's rate of rotation. *J. Coastal Res.*, 8. 966-971.

Mörner, N.-A. (1993a). Global change: the last millennium. *Global Planetary Change*, 7, 211-217.

Mörner, N.-A. (1993b). Global change: the high-amplitude changes 13-10 ka ago – novel aspects. *Global Planetary Change*, 7, 243-250.

Mörner, N.-A. (1995a). Sea Level and Climate – The decadal-to-century signals. *J. Coastal Res.* Sp. I. 17, 261-268.

Mörner, N.-A. (1995b). Earth rotation, ocean circulation and paleoclimate. *GeoJournal*, 37, 419-430.

Mörner, N.-A. (1996a). Global Change and interaction of Earth rotation, ocean circulation and paleoclimate. *An. Brazilian Acad. Sc.* (1996) 68 (Supl. 1), 77-94.

Mörner, N.-A. (1996b). Earth Rotation, Ocean Circulation and Paleoclimate: The North Atlanic–European case. In: *Late Quateranry Palaeoceanography of the North Atlantic Margins* (J.T. Andrews, W.E.N. Austin, H. Bergsten and A.E. Jennings, eds), *Geol. Soc. Spec. Publ.* 111, 359-370.

Mörner, N.-A. (2004). Estimating future sea level changes. *Global Planet.Change*, 40, 49-54.

Mörner, N.-A. (2010). Solar Minima, Earth's Rotation and Little Ice Ages in the Past and in the Future. The North Atlantic – European case. *Global Planetary Change*, 72, 282-293.

Mörner, N.-A. (2011). Arctic environment by the middle of this century. *Energy and Environment*, 22 (3), 207-218.

Mörner, N.-A. (2012). Solar Wind, Earth's rotation and changes in terrestrial climate. *Submitted.*

Mörner, N.-A., Nevanlinna, H. and Shumilov, O. (2003). Past paleoclimatic changes, origin and prediction. *EGS-AUG-EUG meeting*, Nice, April 6-11, Abstracts CL2.07.

Mörth, H.T. and Schlamminger, L. (1979). Planetary motion, sunspots and climate. In: *Solar-terrestrial influence on weather and climate* (B.M. MaCormac and T.A. Seliga, Eds), p. 193-207. Reidel.

Mufti, S. and Shah, G.N. (2011). Solar-geomagnetic activity influence on Earth's climate. *J. Atmospheric Solar-Terrestrial Physics*, 73, 1607-1615.

Patterson, R.T., Prokoph, A. and Chang, A. (2004). Late Holocene sedimentary response to solar and cosmic ray activity influenced climate variability in the NE Pacific. *Sedimentary Geology*, 172, 67-84.

Rochester, M.G. (1984). Causes of fluctuations in the rotation of the Earth. *Phil. Trans. Roy. Soc. London*, A, 313, 95 105.

Roberts, P.H., Yu, Z.J. and Russell, C.T. (2007). On the 60-year signal from the core. *Geophysical and Astrophysical Fluid Dynamics*, 101, 11-35.

Rosen, R.D. and Salstein, D.A. (2000). Multidecadal signals in the interannual variability of atmospheric angular momentum. *Clim. Dyn.* 16, 693-700.

Scafetta, N. (2010). Empirical evidence for a celestial origin of the climate oscillations and its implications. *J. Atmospheric and Solar-Terrestrial Physics*, 72, 951-970.

Scafetta, N. (2011). Total solar irradiance satellite composites and their phenomenological effect on climate. In: *Evidenced-based Climate Science* (D.J. Easterbrook, Ed.), p. 289-316, Elsevier.

Scafetta, N. (2012). Testing an astronomically based decadal-scale empirical harmonic climate model versus the IPCC (2007) general circulation climate models. *J. Atmospheric and Solar-Terrestrial Physics*, in press.

Stuiver, M. and Quay, P.D., 1980. Changes in atmospheric carbon-14 attributed to a variable Sun. *Science* 270, 11-19.

Svensmark, H. (1998). Influence of Cosmic Rays on Earth's Climate. *Physical Review Letters* 81: 5027–5030.

Svensmark, H. (2007). Cosmoclimatology: A new theory emerges. *Astronomy and Geophysics* 48 (1): 1.18–1.24. doi:10.1111/j.1468-4004.2007. 48118.x.

Svensmark, H. and Calder, N. (2007). *The Chilling Stars: A New Theory of Climate Change*. Totem Books. ISBN 978-1840468151.

Thompson, D. W. J. and Wallace, J.M. (1998). The Arctic oscillation signature in the wintertime geopotential height and temperature fields. *Geophysical Research Letters*, 25 (9), 1297–1300.

Willson, R.C. (1997). Total solar irradiance trend during solar cycles 21 and 22. *Science* 277, 1963-1965.

Reviewed by: Professor Don J. Easterbrook
Professor emeritus in geology at Western Washington University, Bellingham, USA. dbunny14@yahoo.com

In: Solar Wind ISBN 978-1-62081-979-1
Editors: C. D. E. Borrega, A. F. B. Cruz

Chapter 3

WAS THE IDEA OF SOLAR WIND WAFTING AROUND EVEN BEFORE THE PARKER FORMULATION?

R. P. Kane*

Instituto Nacional de Pesquisas Espaciais – INPE,
São Jose´ dos Campos, SP, Brazil

ABSTRACT

The idea of solar wind was conceived and formulated by Eugune Parker in a rigorous way in mid 1950s. However, it seems that the idea was floating around in scientific circles even before in some form or other.

Keywords: Solar wind; Parker formulation

1. INTRODUCTION

In the early 1950s, Biermann (1951) proposed that the pointing of comet tails always away from the Sun implies plasma flying out of the Sun on a continuous basis. Since his papers were in German and had a very limited

* E-mail: kane@dge.inpe.br

readership in post-war Germany, Biermann wanted to seek international recognition. So in 1954, he went to USA and met Dr. John Simpson, a heavyweight in American science, at the University of Chicago. Even after prolonged discussions, Simpson was not convinced and rejected the Biermann idea on the grounds that if such continuous emissions occurred, the Sun would be exhausted in no time and would not have lasted for billions of years. Biermann went back, greatly disappointed; but the discussion remained in the back of the mind of Simpson. So, when Eugene Parker came to Simpson´s group in the University of Chicago in 1956, Simpson put this problem in Parker´s lap and goaded him to prove that Biermann was wrong, on the grounds that the Sun, according to Chapman, had a static atmosphere like the Earth, only with much larger dimensions, and the Sun could not afford to emit such emissions continuously. After an intense study of a few months, Eugene Parker, a fresh Ph. D from Caltech who had his thesis on Astrophysical plasma and dust, came to an altogether different conclusion and proved that Chapman was wrong.

The Sun had a highly dynamic atmosphere; and, with temperatures of millions of degrees, the solar corona would be boiling and would emit corpuscular radiation *continuously*, sometimes more, sometimes less but never zero. This continuous emission could be termed as "solar wind". Parker wrote papers on this topic (Parker 1958, 1959) which were mostly ignored or despised by the scientific community (e.g., Chamberlain, 1960). Nevertheless, Parker continued his investigation and predicted that the quiet time solar wind would stretch the solar dipole magnetic field in the interplanetary space. Thus, interplanetary space could no more be considered as a vacuum but was filled with plasma entwined in spiral magnetic field structures. A new discipline, namely Heliospheric Physics, was born. Parker had a valuable companion in Thomas Gold from Harvard Observatory, who extended Parker´s ideas to abnormal solar phenomena and predicted that fast solar wind associated with solar flare emissions would result in interplanetary magnetic bottles, which, on reaching the Earth´s orbit at 1 AU, would cause geomagnetic storms if the Earth happened to be in the path. Gold (1959) also coined the term "magnetosphere", in which the solar wind would confine Earth´s normally dipole magnetic field, compressed on the sunward side and stretched as a long tail in the anti-sunward direction. In spite of the strong opposition, Parker and Gold pursued their ideas vehemently and were finally vindicated by satellite observations, first by Gringauz (1961) and soon after by Bridge et. al (1962) and Neugebauer and Snyder (1962). Parker became famous as a scientific hero overnight.

2. EARLIER OBSERVATIONS

An examination of earlier history seems to indicate that ideas similar to Parker´s solar wind were floating around during several earlier decades. Biermann (1951) used the term *"solare Korpuskularstrahlung" meaning solar corpuscular radiation, but there was no mention of the term "wind" as such.* Veselovsky (2009a) mentions Kiepenheuer´s contribution, where. Kiepenheuer (1953, p.440) states, "coronal streamers and the straight comet tails (type 1) have the same cause, a `wind´ blowing from the Sun". So, Kiepenheuer was probably the first, who coined the term "wind" in his classical review description of the nearly radial permanent flow of plasma everywhere around the Sun. At that time, he justified and defended physically correct orders of magnitudes for main parameters: plasma density, velocity and temperature together with their variations based on indirect data. Numerous observations of coronal rays, comets, geomagnetic storms, solar eruptions and cosmic rays were consistently interpreted on this basis.

Far back in time, it all seems to have started from Richard Christopher Carrington (26 May 1826 – 27 November 1875), who was an English amateur astronomer, whose 1859 astronomical observations first corroborated the existence of solar flares as well as their electrical influence upon the Earth and its aurorae; and whose 1863 records of sunspot observations demonstrated differential rotation in the Sun. In recent decades, Chapman and Bartels (1940) specifically mentioned "solar corpuscular radiation" as the source of geomagnetic storms which occurred roughly 20-100 hours after the eruption of solar flares. Earlier, Chapman (1929) correctly calculated the Archimedian spiral shape of the corpuscular streams from the rotating Sun responsible for the recurrent and sporadic geomagnetic storms. But he was wrong with his *a priori* assumption about a small angular width of those streams. In reality, they are not 10 degrees and less, but much broader and are a global phenomena. It is curious to note that the concept of the static corona dominated, which is totally in error at large distances from the Sun. On the other hand, the useful concept of the magnetosphere (not actually the word magnetosphere) was initially introduced by. Chapman and J. Bartels (1940) as if it was a purely transient phenomenon, rather the opposite to what we know now.

In the opinion of Veselovsky (2009a), the question: "Who first discovered the solar wind?" (Schröder, 2008) seems to be ill posed, as it has no definite answer. Attempts to prescribe this 'discovery' to one or few individuals are not justified by the long history of the relevant studies. There is a list of important contributors and we should carefully keep our memory of them.

However, since Eugene Parker was a fresh Ph. D. when Simpson persuaded him to look into this problem and Parker made rigorous calculations in a few months and concluded that a "continuous" outflow from the Sun was necessary to balance his equations, he probably coined the term "solar wind" by himself, unaware of previous literature.

Unfortunately, the permanent and global character of the contemporary solar wind was neglected or not correctly represented even in orders of magnitudes (being, especially, too small or too high for the density) by many other authors until the space era, when the first direct measurements put an end to such prejudices. Parker´s formulation was much more realistic.

Some parts of this topic were presented in Kane (2006, 2009a) while Veselovsky (2009 a,b,c) discussed the origin of solar wind. Incidentally, the expansion of corona was studied by Bondi (1952) who surmised "symmetric" accretion of the corona, inward as well as outward. Parker only dealt with the outward flow. The inward flow probably exists; but when super hot coronal plasma invades the chromosphere, the comparatively dense chromosphere resists violently. Some heat percolates down below and the chromosphere temperature rises to ~10,000 K, higher than the photosphere temperature, ~6,000K. But the main consequence is the formation of a thin (few hundred kilometrers) "transition region", where the temperatures are ~100,000K, higher than the chromospheric temperatures (10,000K) but much lower than the coronal temperatures (a few million K). In high sunspot activity when coronal temperatures increase further, the transition region is pushed down from ~2,100 km in quiet Sun, to ~1,700 km at higher sunspot activity (e. g., see Figure 1 in Kane, 2009b).

ACKNOWLEDGMENT

This work was partially supported by FNDCT, Brazil, under contract FINEP-537/CT.

REFERENCES

Biermann, L., Kometschweife und solare Korpuskularstrahlung. *Z. Astrophys.* 29, 274–286, 1951.

Bondi, H., On Spherically Symmetric Accretion, *Mon. Not. R. Astron. Soc.* 112, 195–204, 1952.

Bridge, H.S., Dilworth, C., Lazarus, J., Lyon, E.F., Rossi, B., Scherb, F., Direct observations of the interplanetary plasma. *J. Phys. Soc. Jpn.* 17, (Suppl. A-II), 553–559, 1962.

Chamberlain, J.W., Interplanetary gas: II. Expansion of a model corona. *Astrophys. J.* 131, 47–56, 1960.

Chapman, S., Solar streams of corpuscles: their geometry, absorption of light, and penetration. *Mon. Not. Roy. Astrono. Soc.,* 89, #3, 456-470, 1929.

Chapman S. and Bartels, J., *Geomagnetism,* vol. 2, Oxford University Press, London, *1940.*

Gold, T., Plasma and magnetic fields in the solar system. *J. Geophys. Res.* 64, 1665–1674, 1959.

Gringauz, K.I., Bezrukikh, V.V., Ozerov, V.D., Rybchinskii, R.E., A study of the interplanetary ionized gas, high-energy electrons, and corpuscular radiation from the sun by means of the three-electrode trap for charged particles on the second Soviet cosmic rocket. *Soviet Phys. Dokl.* 5, 361–364, 1960.

Kane, R.P., The idea of space weather – a historical perspective. *Adv. Space Res.* 37, 1261–1264, 2006.

Kane, R. P., Early history of cosmic rays and solar wind – Some personal remembrances, *Adv. Space Res.,* 44, 1252–1255, 2009a.

Kane R. P., Gnevyshev peaks in solar radio emissions at different frequencies, *Annales. Geophys.,* 27, 1469-1475, 2009b.

Kiepenheuer, K. O., Solar activity. Chapter VI. In: *The Sun* (Ed.G. P. Kuiper). Chicago: University of Chicago Press, pp 322-465, 1953.

Neugebauer, M.M. and, Snyder, C.W., Solar plasma experiment., *Science.* 138, 1095–1096, 1962.

Parker, E.N., Suprathermal particle generation in the solar corona. *Astrophys. J.* 128, 677–685, 1958.

Parker, E.N., Extension of the solar corona into interplanetary space. *J. Geophys. Res.* 64, 1675–1681, 1959.

Schröder, W., Who first discovered the Solar wind? *Space Reaserch Today,* 171, 34-35, 2008.

Veselovsky, I. S., When and how the solar wind started to blow? *Space Research Today,* #174, 67-68, 2009a.

Veselovsky, I. S., Origin of the Solar Wind: Astrophysical and Plasma_Physical Aspects of the Problem, ISSN 0016_7932,

Geomagnetism and Aeronomy, 2009, Vol. 49, No. 8, pp. 1148–1153. © Pleiades Publishing, Ltd., 2009b.

Original Russian Text © I.S. Veselovsky, 2008, published in *Solnechno_Zemnaya Fizika,* 2008, Vol. 12, No. 1, pp. 93–98.

Veselovsky, I. S., Solar Wind origins and statistical properties, Paper presented at the IAGA International Symposium-2, Cairo, Egypt, Dec. 4-8, 2009, Abstract Book p. 45, 2009c.

In: Solar Wind ISBN 978-1-62081-979-1
Editors: C. D. E. Borrega, A. F. B. Cruz

Chapter 4

SOLAR WIND INFLUENCE ON ATMOSPHERIC PROCESSES IN WINTER ANTARCTICA

O. A. Troshichev* [*], *V. Ya. Vovk and L. V. Egorova
Arctic and Antarctic Institute, Russia

ABSTRACT

The paper presents a summary of the experimental results demonstrating the strong influence of the interplanetary electric field on atmospheric processes in the central Antarctica, where the large-scale system of vertical circulation is formed during the winter seasons.

The influence is realized through acceleration of the air masses, descending into the lower atmosphere from the troposphere, and formation of cloudiness above the Antarctic Ridge, where the descending air masses enter the surface layer.

The cloudiness formation results in the sudden warmings in the surface atmosphere, since the cloud layer efficiently backscatters the long wavelength radiation from the ice sheet, but does not affect the adiabatic warming process of the descending tropospheric air masses.

The acceleration is followed by a sharp increase of the atmospheric pressure in the near-pole region, which gives rise to the katabatic wind strengthening above the entire Antarctica.

---

[*] Eg225. Arctic and Antarctic Research Institute, St. Petersburg, 199397, Russia. Fax:+7-812-352-2688, Phone: 7-812-337-3134. E-mail: olegtro@aari.nw.ru.

As a result, the circumpolar vortex about the periphery of the Antarctic continent correspondingly decays and the cold air masses flow out to the Southern ocean. The latter phenomena evidently destroys the regular relationships between the sea level pressure fluctuations in the Southeast Pacific high and the North Australian-Indonesian low. It seems that the El-Niño beginnings are related exactly to the anomalous atmospheric processes in the winter Antarctica.

Keywords: Solar wind, Antarctica, atmosphere, anomalous winds, El-Niño

INTRODUCTION

Existing models of the atmospheric variability and change do not take into consideration the short-term changes of solar activity. Indeed, the total energy, contributed by the solar wind and the cosmic rays in the Earth's atmosphere, is extremely insignificant in comparison with the total solar irradiance. But, as distinct from the total solar irradiance, the energy of solar wind and cosmic rays can increase in hundreds and more times in periods of high solar activity.

The attempts to find the cause-effect relations between the solar activity variations and weather and climate changeability have a long story [Wilcox, 1975; Herman and Goldberg, 1978].

The galactic cosmic rays (GCR) altered by solar wind were usually regarded as the most plausible agent of the solar activity influence on the Earth's atmosphere. The experimental data were presented showing the influence of the varying GCR flux on the Earth's weather and climate [Tinsley et al., 1989], on high cloud coverage [Pudovkin and Veretenenko, 1995], on temperature in the polar troposphere [Pudovkin et al., 1996, 1997], on the global total cloud cover [Svensmark and Friis-Christensen, 1997; Todd and Kniveton, 2001], on low cloud coverage [Marsh and Svensmark, 2003].

These results suggest that just cloudiness variation affected by cosmic rays lead to changes in the atmospheric and meteorological characteristics.

However, the hypothesis about the determining influence of the galactic cosmic rays on the cloudiness was not always supported by the subsequent, more detail research. It was indicated that the correlation with GCR disappears when the cloud coverage is decomposed in fractions by cloud type or height, by region (reduce for ocean basis), or by latitude (patterns in the tropical zone are better associated with concurrent El-Nino) [Farrar, 2000].

A comprehensive study of the low cloud coverage for the last 120 years [Palle and Butler, 2002] revealed that the global cloudiness increased during the past century regardless of variations of GCR. The solar irradiance turned out to be correlated better and more consistently with low cloud cover than the cosmic ray flux [Kristjansson et al., 2002]. As a result, the conclusion was maid that the mechanism linking the cosmic ray ionization and cloud properties cannot be excluded, but its high efficiency is not obvious [Harrison and Carslaw, 2003].

At the same time it is well known that severe reductions in the galactic cosmic rays flux, known as Forbush decrease (FD), are related to the disturbed, high speed solar wind, emerging by the most intense solar flares. The disturbed solar wind is characterized by the largest changes in such parameters as the solar wind pressure $P_{SW} = m \cdot n \cdot (V_{SW})^2$ and the interplanetary electric field $E_{SW} = V_{SW} \times B_Z$ (or simply the southward component of the interplanetary magnetic field (IMF B_{ZS}). The velocity V_{SW} might vary with a factor of 2-3 at maximum, whereas the IMF B_Z components might change the sign and increase by some factors of ten. The geoeffective solar wind parameters strongly affect (impact) the Earth's magnetosphere. It was noted that the FD beginnings at the Earth's orbit are recorded simultaneously with dramatic disturbances in the solar wind, and therefore, the atmospheric effects, assigned to Forbush decreases, can be, in reality, influenced by the geoeffective solar wind parameters [Troshichev et al., 2003].

Figure 1. shows, as an example, response of the cloudiness above Vostok station (Antarctica) to Forbush decrease (left column) and to the IMF B_Z minimum (right column) during the winter season of 1974-1992 [Troshichev et al., 2008]. The list of 24 Forbush decreases was taken from the widely known analysis [Todd and Kniveton, 2001], but in our case the Forbush decrease maximum has been used as a key date in the epoch superposition method unlike to Todd and Kniveton [2001], who used the Forbush decrease beginning. Indeed, in many cases it is difficult to determine the FD beginning unambiguously, as a result the FD beginning dates sometimes are identified with as large a scatter as 5 days in various studies. On contrast, the Forbush event maximum is easily and uniquely identified by minimum in the galactic cosmic ray flux in each case.

The same list has been used to separate the IMF B_Z minimum dates, related to the FD events. Unfortunately, the IMF B_Z data turned out to be available only for 15 events of 24. So, the left column in Figure 1. is for the data of cloudiness above Vostok, allocated relative to the FD maximum, whereas the right column is for the data allocated to the appropriate IMF B_Z minimum.

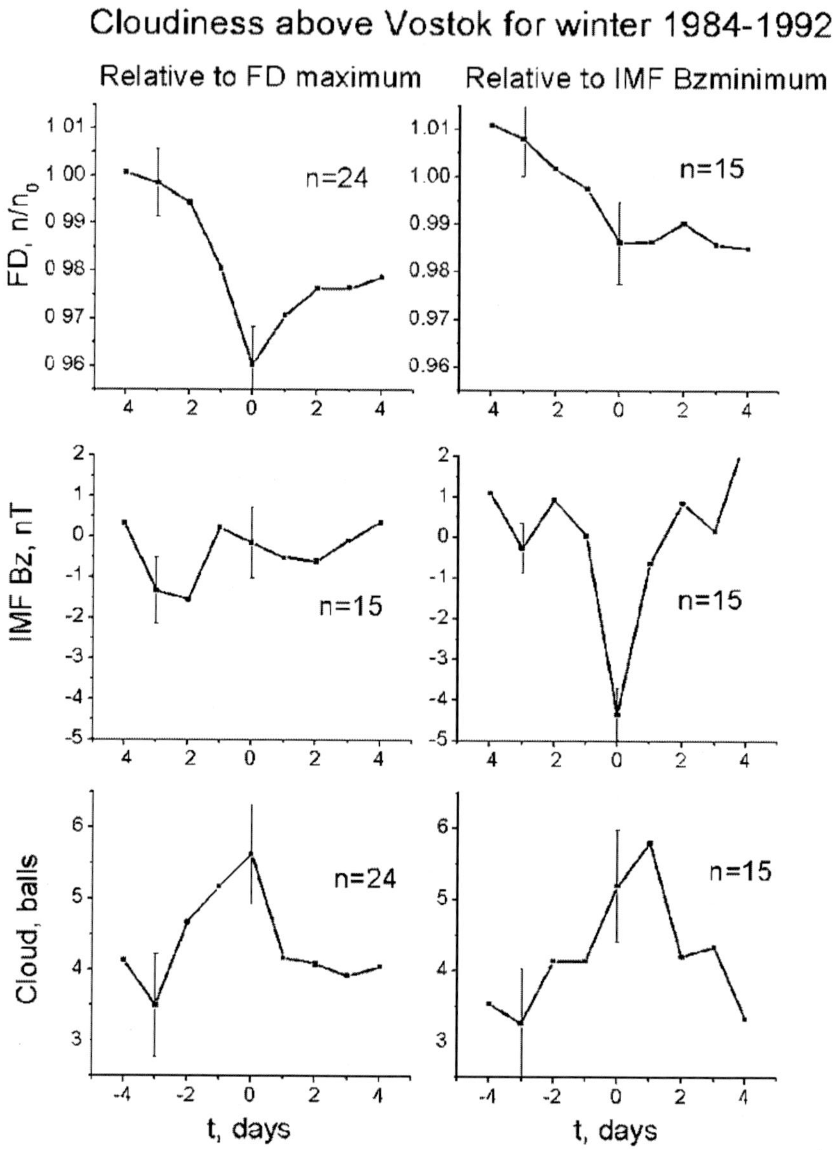

Cloudiness above Vostok for winter 1984-1992

Relative to FD maximum Relative to IMF Bzminimum

Figure 1. Behavior of the average Forbush decrease (FD), the average interplanetary magnetic field (IMF) B_{ZS} component, and the appropriate cloudiness above Vostok for the most powerful FD events during the winter season of 1974-1992 (list of Todd and Kniveton, [2001]). A key date (t = 0) was taken as a day with FD maximum in the left column and a day with the IMF B_Z minimum in the right column [from Troshichev et al.,2008].

The results presented in the left column demonstrate that Forbush decrease coincides with increased cloudiness, which starts three days ahead of the key date (FD minimum) and reaches the 55 % maximum by the key date, the statistic significance being equal to 0.96. The results presented in the right column demonstrate, with no less evidence, that cloudiness above Vostok starts to increase one day before the IMF B_Z minimum and reaches its maximum next day after the key date. The statistical significance in this case is less, ss=0.91, but we have to take into account that number of the available events was reduced in 1.5 times while examining the Bz indicator instead of Forbush decrease. This example clearly shows that changes of cloudiness may be successfully explained by Forbush decrease as well as by the IMF variations.

IMF Variations as a Determining Factor for the Cloudiness above Vostok

To demonstrate that variations in the interplanetary magnetic field by themselves can produce an effect on cloudiness we need to examine such solar wind disturbances, which were accompanied by the quite insignificant Forbush decreases. Taking into account that the Forbush decrease magnitude is negligible for the solar minimum epochs, the relation between the interplanetary magnetic field and the cloudiness above Vostok was examined for the years of the solar minimum (1974-1977 and 1985-1987) [Troshichev et al., 2008]. The cloudiness at Vostok station was determined by two methods. The first method is based on estimation of cloudiness power in reports from the visual man-made observations (0 is for clear sky, 10 is for heavy cloudiness). The second method is based on measurement of the radiation balance (BR) value in MJ/m^2 produced by balancer. It has been known that during the winter season, under conditions of the dark polar night, the radiation balance at Vostok is always negative. The larger negative BR values correspond to more intense radiation cooling, the less negative BR values indicate the cooling reduction as a consequence of the cloud layer formation above Vostok station.

Figure 2. demonstrates the response of the radiation balance (second panel), the cloudiness balls (third panel) to the negative deviation in the daily averaged IMF Bz component (top panel) for three groups of B_Z values: $-2 < B_Z < -1$ nT (18 events), $-2.5 < B_Z < -2$ nT (11 events), and $B_Z < -2.5$ nT

(13 events) in years of solar minimum (1974-1977, 1985-1987), the day of maximal negative B_Z deviation being taken as a zero date (t = 0). One can see the evident response of cloudiness to influence of the interplanetary magnetic field: the greater the negative IMF B_Z component, the larger is the cloudiness, the more pronounced is the reduction in the cooling.

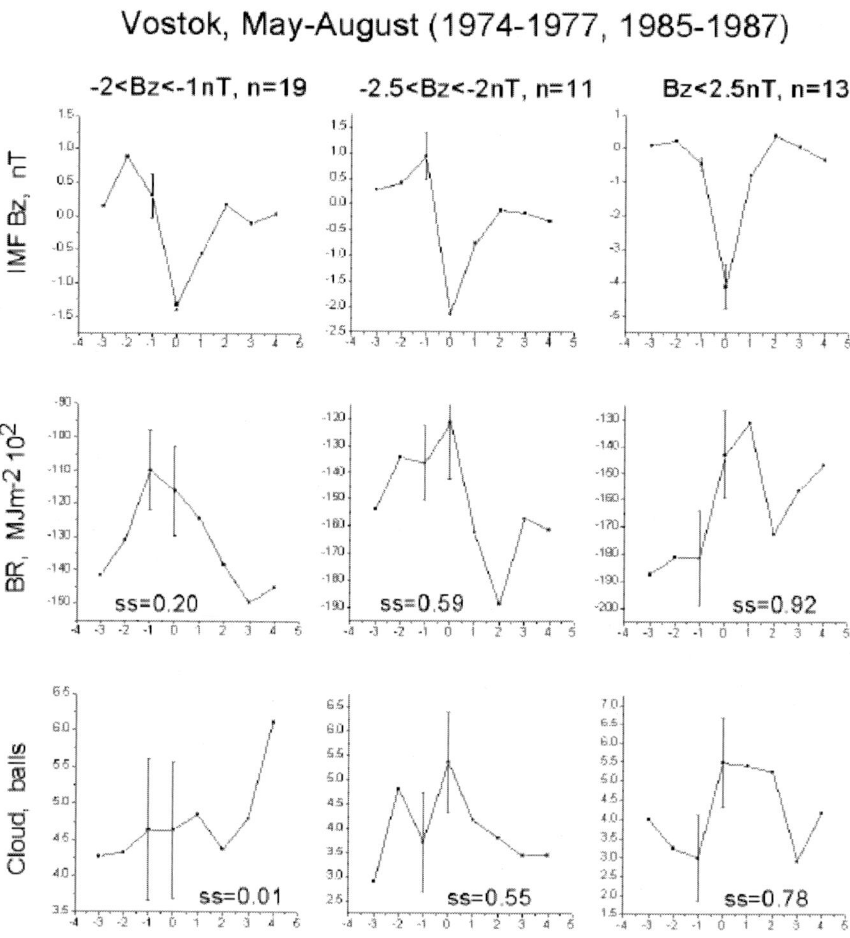

Figure 2. Mean changes in cloudiness estimated from the radiation balance measurements (second panel) and by the visual man-made observations (third panel), obtained for three gradations of the negative deviation in the daily averaged IMF B_Z component ($-2<B_Z<-1$ nT, $-2.5<B_Z<-2$ nT, $B_Z<-2.5$ nT) during the winter season of 1974-1977 and 1985-1987 [from Troshichev et al., 2008].

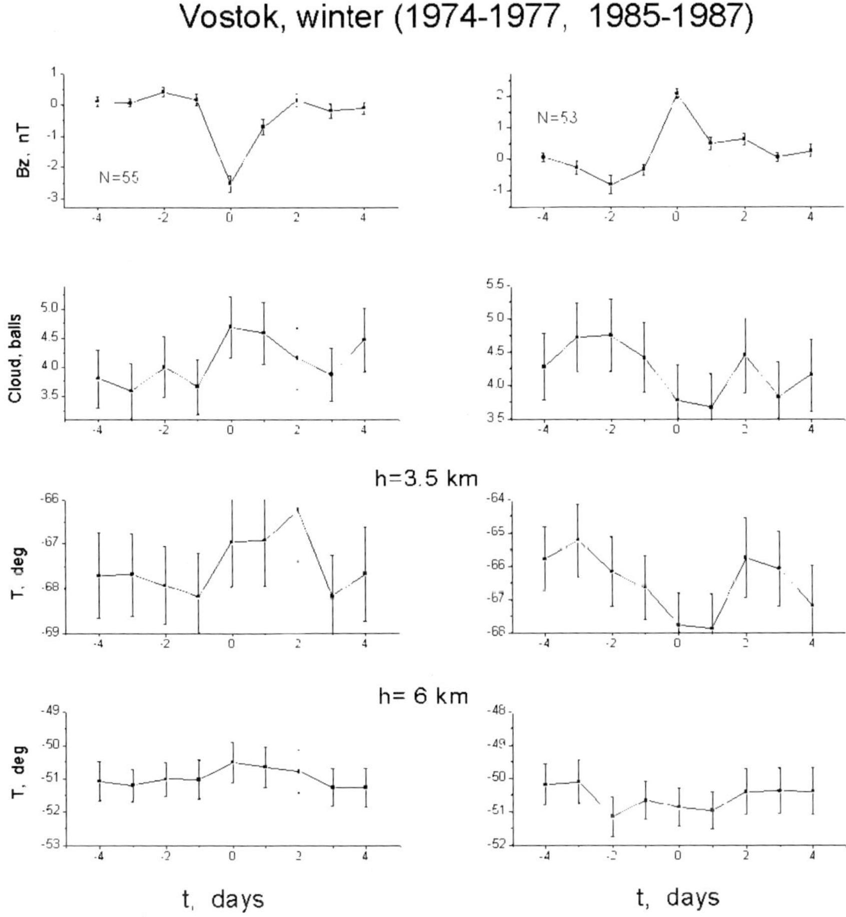

Figure 3. Run of the daily averaged IMF B_Z component for cases of strong negative and positive B_Z deviations (upper panel) and the appropriate effects in the cloudiness (second panel) and in temperature at heights 3.5 km and 6 km above Vostok (third and forth panels) in years of the solar activity minimum 1974-1977 and 1985-1987 [from Troshichev et al., 2008].

The cloudiness formation starts simultaneously with the negative B_Z deviation (-1st day) and reaches the maximum at 0 or +1st day. It is important that statistical significance of all effects, being minor for the first B_Z gradation, quickly grows with the increase of the negative B_Z (in spite of the events number diminishing) and reaches 92% level in case of radiation balance for third B_Z level.

The crucial role of the interplanetary magnetic field was confirmed when examining the effects of the strong negative (ΔB_Z<-2 nT) and positive (ΔB_Z>2 nT) deviations of the IMF B_Z component [Troshichev et al., 2008]. As Figure 3. shows the intensification of the negative B_Z deviation is followed, with a delay time of about 1 day, by the cloudiness enhancement and the corresponding warming in the ground layer, whereas the rise of the positive B_Z deviation is followed by the opposite reaction: the cloudiness decays and the cooling starts on the ground layer. It should be reminded that Vostok is located at the ice dome at an altitude of 3.45 km above sea level, and, therefore, h=3.5 km corresponds to ground level at Vostok, whereas h= 6 km corresponds to an altitude of ~ 2.5 km above the ice sheet level.

The results of the analysis [Troshichev et al., 2008], completed for conditions of the negligible Forbush decrease, demonstrate that the sign of the B_Z component defines the trend of the cloudiness change (growth or decay), the cloudiness power being determined by the value of the southward IMF component. It seems reasonable to suggest that the formation of the cloud layer occurs at altitudes higher than ~ 5 km above the ice sheet.

A Distinctive Features of the Atmospheric Circulation above Antarctica

The Antarctic continent is dominated by the ice dome rising above 3.5 km in the Antarctic ridge area. A unique feature of the atmospheric circulation in the winter Antarctica is the continental-scale katabatic wind regime [Egger, 1985; Parish and Bromwich, 1987, 1991]. It is a powerful drainage stream of the near-surface air masses flowing roughly radially from the Antarctic ridge to coastline (Figure 4). This drainage is determined by the negative air buoyancy supported by severe radiation cooling of the atmosphere on the ice sheet surface. Since mass continuity requires a permanent substitution of the air masses draining in the near-surface layer, the air masses are supplied from the troposphere over the Antarctic ridge area. As a result, a large-scale system of vertical (meridional) circulation is formed in the winter Antarctica (Figure 5). The system includes drainage of the air masses along the slope of the Antarctic ice sheet, ascending flow near the coast line, return movement in the lower and middle troposphere, and descending flow in the near-pole region [Parish and Bromwich, 1991].

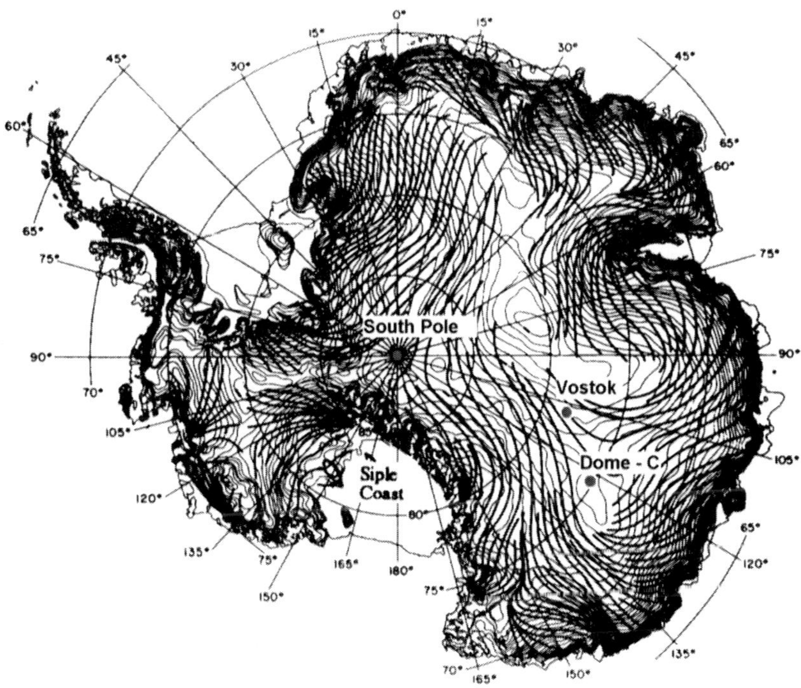

Figure 4. Drainage pattern of near-surface katabatic winds [from Parish and Bromvich, 1991]. Location of the inner-continental stations Vostok, Dome C and South Pole is shown.

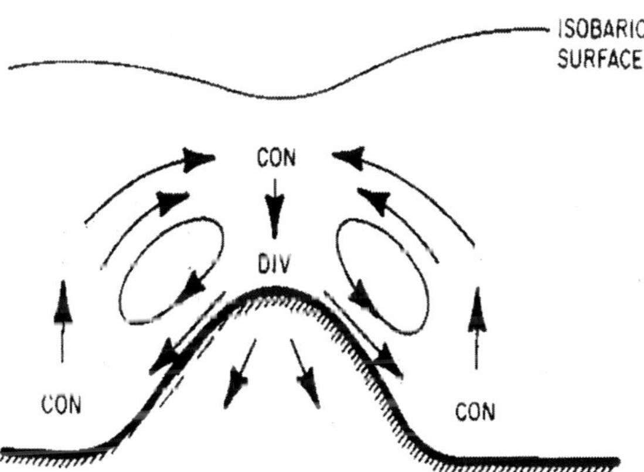

Figure 5. A conceptual scheme of the vertical mass circulation forced by the katabatic wind regime in Antarctica [from Parish and Bromwich, 1991].

The spatial structure of katabatic winds is one of the most stable atmospheric phenomena on the Earth [Schwerdtfeger, 1984]. Phases of weak and strong katabatic winds alternate with each other at regular intervals from few days [Egger, 1985; James, 1989] to 30-50 days [Yasunari and Kodama, 1993]. The model-based studies [Egger, 1985; James, 1989; Parish and Bromvich, 1991; Parish, 1992] showed that the Antarctic katabatic wind regime appears to be an important forcing mechanism for the circumpolar vortex around the periphery of the Antarctic continent.

As the experimental study of Yasunari and Kodama [1993] demonstrates, the weak katabatic wind phase corresponds to a deep upper tropospheric circumpolar vortex, whereas the strong katabatic wind phase corresponds to a weak circumpolar vortex. Thus, the low atmosphere is not static above the Central Antarctic ridge: there is a powerful vertical channel, where the air masses go down from upper troposphere to the Antarctic ice sheet.

SUDDEN WARMINGS IN THE CENTRAL ANTARCTIC AND THEIR RELATION TO THE DISTURBED SOLAR WIND

It was suggested [Troshichev and Janzhura., 2004] that superposition of the constant radiation cooling of air situated at the ice sheet and adiabatic warming of air masses, which arrive from above, maintains the atmosphere on the Antarctic ridge in the state of thermal quasi-equilibrium. A cloud layer formation at altitude about 5-10 km would efficiently backscatter the long wavelength radiation from the ice sheet, but it will not affect the adiabatic warming process, when the air masses go down from troposphere to the Antarctic ice dome. As a result of the reduction in the radiative cooling, the atmosphere should be warmer below the cloud and give rise to sudden warmings at ground level, which happen sometimes in the Central Antarctica during the winter seasons. The particular events of sudden warmings were studied in [Troshichev et al., 2003; Troshichev and Janzhura, 2004] on the basis of three sets of meteorological data:

1. Daily meteorological observations (temperature, pressure and winds) at the ground level at Vostok station (h =3.45 km) for 1978-1992;
2. Hourly temperature values derived from the 10-min observations provided by the automatic stations (AWS) at Dome C, South Pole and Vostok for 2000-2001; and

3. Daily aerological measurements of temperature, pressure and winds above Vostok station (h =3.5–20 km) for 1978-1992. It was found that warmings came after large increases in the amplitude of the negative (southward B_{ZS}) component of the interplanetary magnetic field. The time of the maximum deviation in the solar wind parameter (B_{ZS}, or E_{SW}, or P_{SW}) was used as a key ("zero") date. The temperature variation (ΔT) was calculated as the difference between the temperature values for the key moment and for the preceding and succeeding days (or hours).

The results (Troshichev et al., 2004) demonstrated the clear correlation between B_Z and ΔT and between E_{SW} and ΔT.

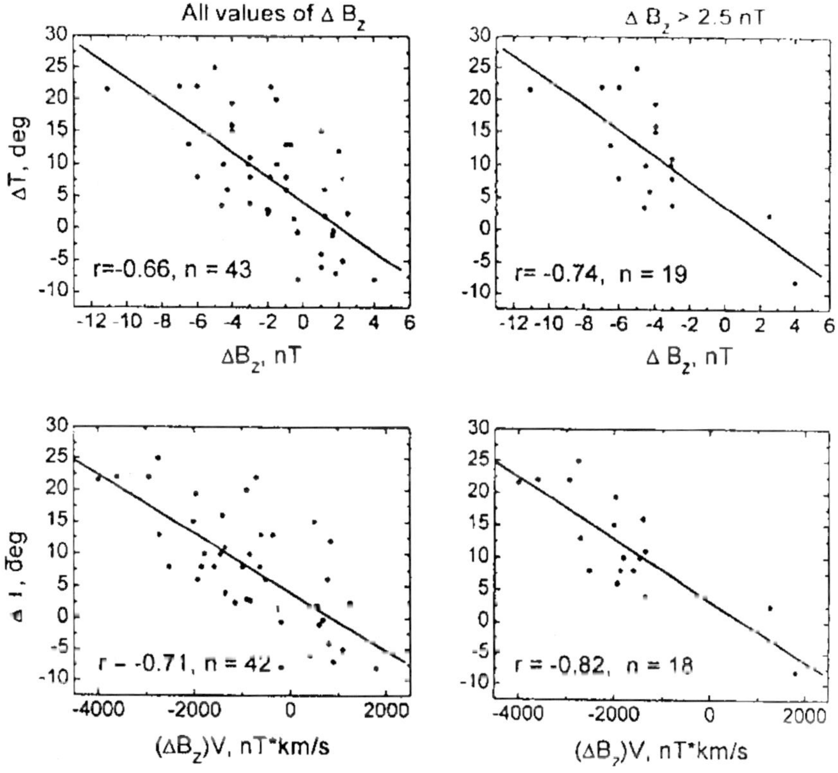

Figure 6. Correlation between daily changes in ground temperature (ΔT) at Vostok station (1978-199) and daily deviations in the interplanetary magnetic field ΔB_Z (upper panel) and in the geoeffective electric ESW field (lower panel) [from Troshichev et al., 2003].

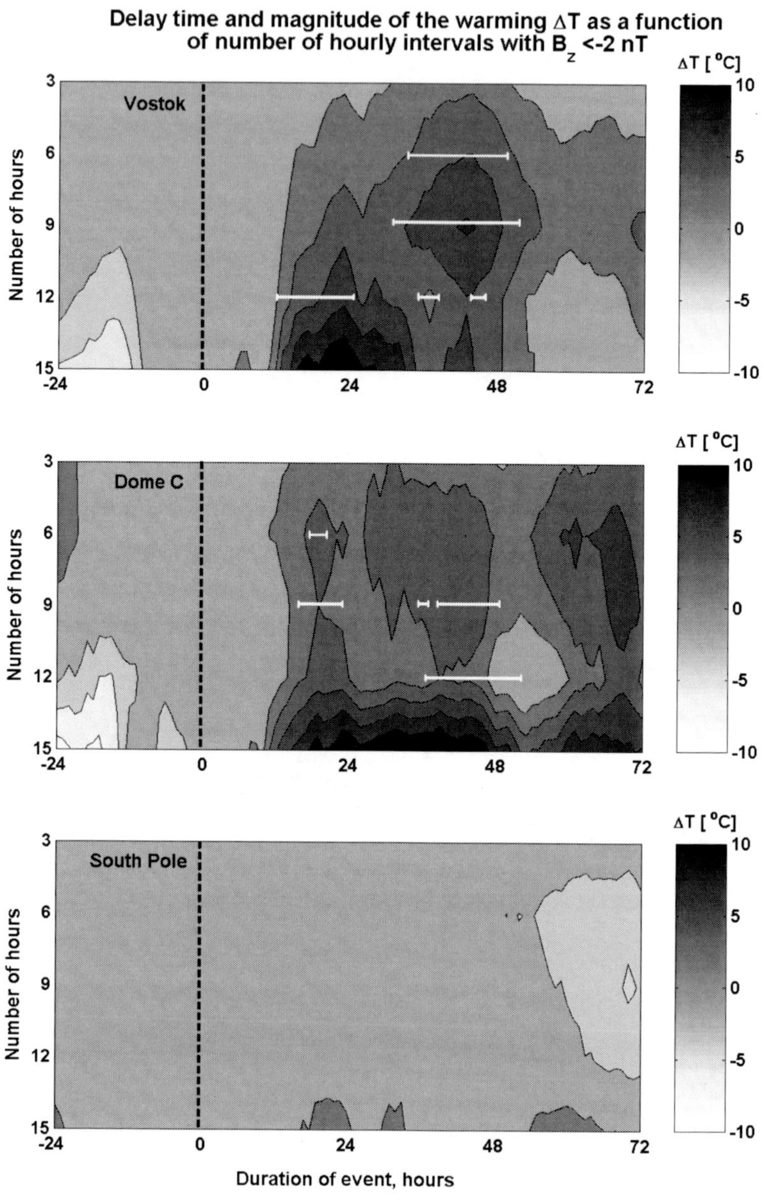

Figure 7. Character of temperature hourly changes ΔT at stations Vostok, Dome C, and South Pole as a function of number of the hourly interval with B_Z<-2nT. Dotted line marks the key moment. Rate of the warming is presented by dark signatures [from Troshichev and Janzhura, 2004].

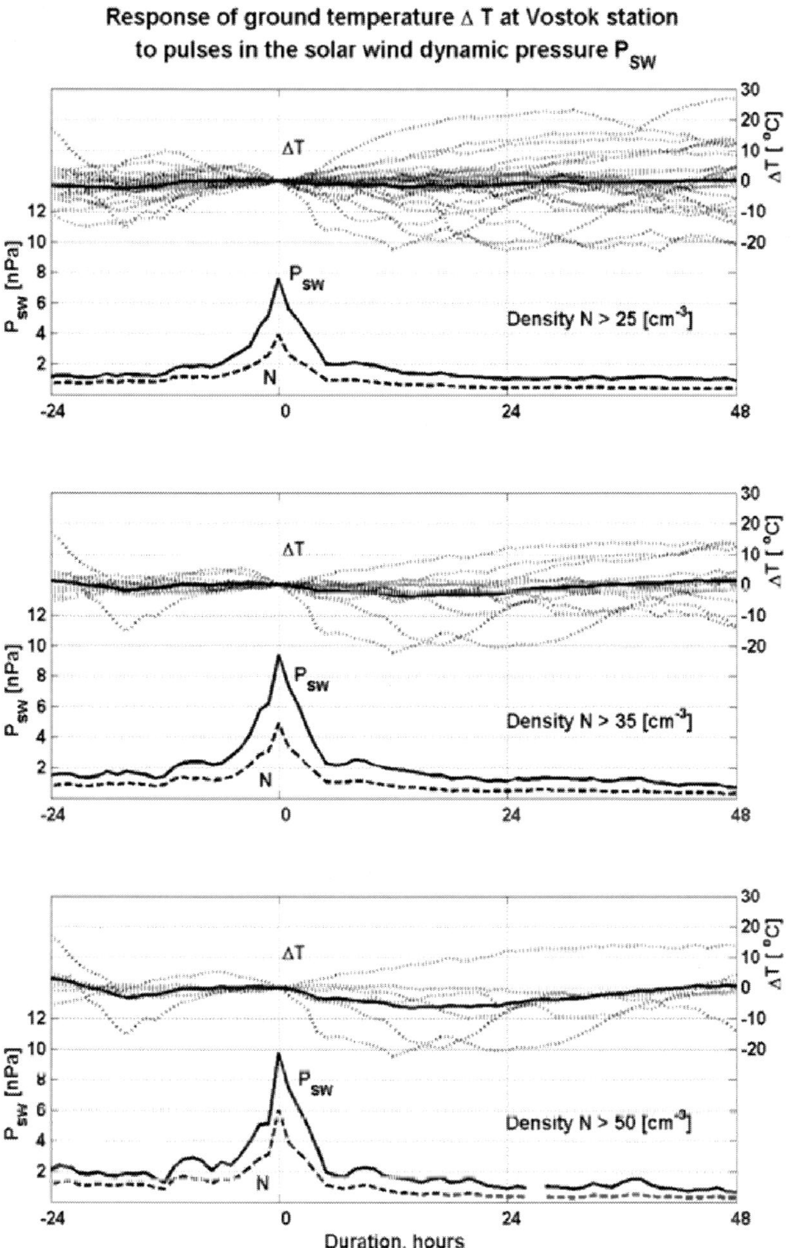

Figure 8. Response of the ground temperature ΔT at Vostok to pulses in the solar wind dynamic pressure in 2000 – 2001, the key moment being determined by sharp increase in the dynamic pressure. [from Troshichev and Janzhura, 2004].

As Figure 6. shows, the correlation of the temperature variations with the electric field deviations ΔE_{SW} is markedly higher than with ΔB_Z and can expressed as:

$$\Delta T\ [deg] = 3.5 - 0.0047 \cdot \Delta E_{SW}\ [nT \cdot km/s],$$

where ΔE_{SW} is regarded as negative for negative values of ΔB_Z. The correlation is best for larger values of the leaps in B_Z, the regression line slope being determined by the end points (maximal negative or positive values ΔB_Z).

Figure 7. shows the statistical response of the ground hourly temperature ΔT at stations Vostok, Dome C and South Pole to variations in E_{SW} in 2000-2001 as a function of the duration of southward IMF [Troshichev and Janzhura, 2004]. The abscissa axis presents the duration of events, the ordinate axis is for the number of successive hourly intervals with negative $B_Z < -2$ nT. One can see that the increase in the ground temperature is determined by the power of the negative B_Z action: the longer the B_{ZS} field exposure (and the higher electric field intensity) lasts, the larger is the temperature deviation and the shorter is the delay time between the key moment and the temperature change. At stations Vostok and Dome C a 15-hours exposure affects the effective warming (up to $\Delta T > +10°$) after $\approx$12 hours at a level of statistical significance equal to 0.99.

However, the warming at station South Pole may be observed only under conditions of very strong interplanetary electric field, and this link is not statistically significant. The different effects in the ground temperature at stations Vostok and Dome C, on the one hand, and South Pole, on the other hand, are easily explained by their different locations relative to the katabatic wind system. Indeed, stations Vostok and Dome C are located on the Antarctic ridge, which is the region of the descending tropospheric air mass flow, whereas South Pole is located in the regions of the developed drainage stream.

As the results [Troshichev and Janzhura, 2004] showed, there is a weak tendency to ground temperature decrease after the pressure pulses: the coolings at Vostok follow the large pressure pulses within 2 hours and last about the next 24 hours (Figure 8). However, we have to keep in mind that pressure pulses in the interplanetary shocks usually pass the Earth some hours (from 0 to 12) ahead of the interplanetary electric field disturbances. So, the isolated effects of the pressure increases can be seen, in principle, only whithin the first hours after the pressure pulses. When the electric field comes into play, the influences of the pressure pulses and interplanetary electric field would superpose on one another, and their combined effect in the temperature deviations is observed.

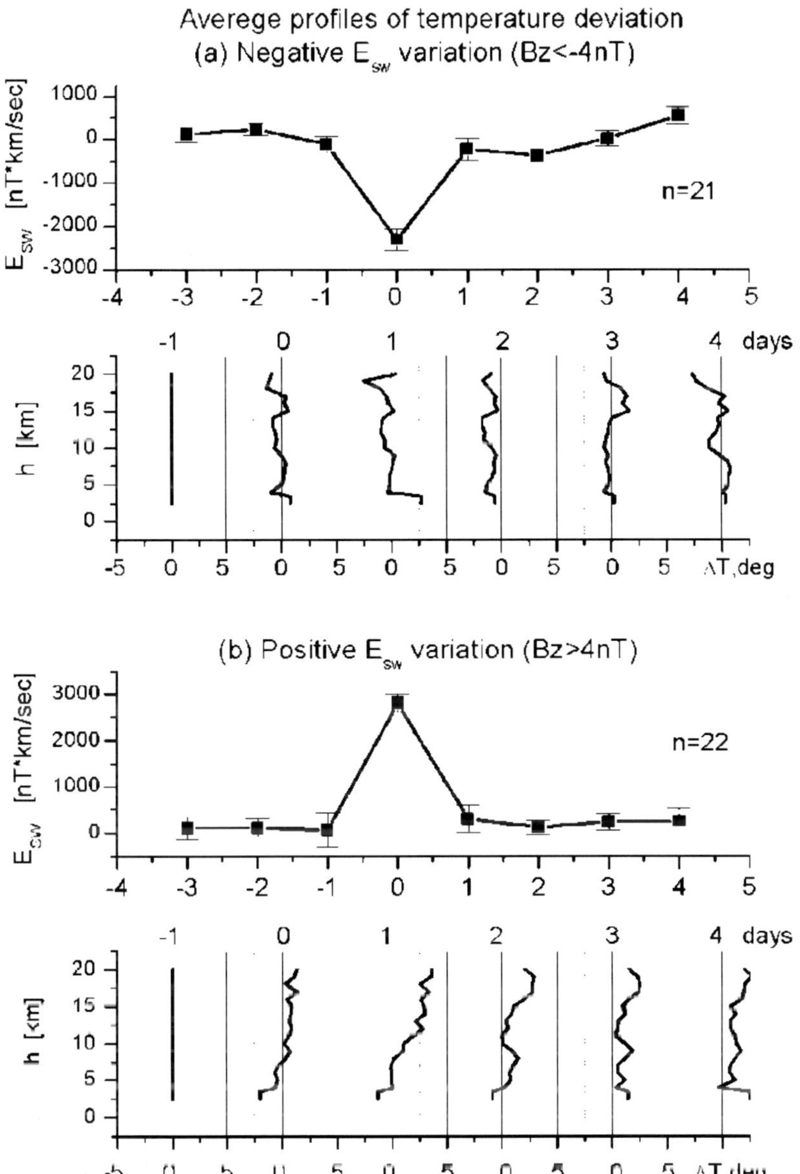

Figure 9. Mean height profiles of the daily temperature deviations above Vostok under conditions of the negative and positive ESW changes [from Troshichev et al., 2003].

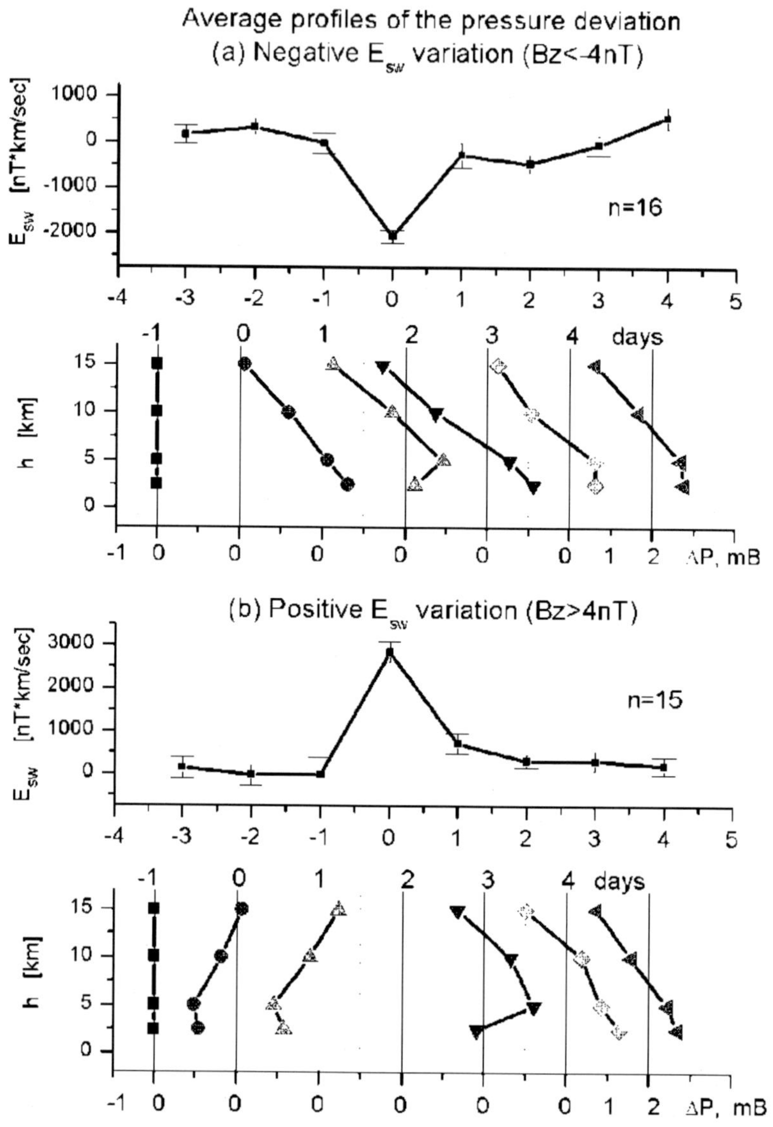

Figure 10. Mean height profiles of the daily atmospheric pressure deviations above Vostok under conditions of the negative and positive E_{SW} changes [from Troshichev et al., 2004].

It is meaningful that the response in temperature to the E_{SW} influence is quite opposite in the lower and upper troposphere. Figure 9. shows the profiles of the averaged daily temperature deviations above Vostok station for negative and positive variations of E_{SW} [Troshichev et al., 2004]. The temperature profile for the –1st day, preceding the zero day (i.e. day of the maximum E_{SW} deviation), is taken as the level of reference for all succeeding days. The average warming at the ground level (h = 3.45-3.5 km) responds, within 1-2 days, to the negative step in ΔE_{SW}, but at altitudes more than 10 km a cooling is observed (upper panels). The opposite behaviour is typical of the positive step in E_{SW}: the atmosphere gets cooler at ground level, and gets warmer at h > 10 km (lower panels). Such regularity would be observed if a cloud layer appears at altitudes 5-10 km under the influence of the negative E_{SW} variations, and disappears under the influence of the positive E_{SW} variations.

Impact of the Disturbed Solar Wind on Atmospheric Pressure

The aerological data from Vostok station for 1978-1992 have been used to study relationship between the daily values of the interplanctary electric field and atmospheric pressure at altitudes h=3.5–20 km [Troshichev et al., 2004]. A day of negative or positive leap in the interplanetary electric field has been taken as a key date in the epoch superposition analysis. The magnitudes of ΔE_{SW} (difference between daily values for the key date and for the preceding and succeeding days) were compared with the appropriate value of the atmosphere pressure. The results presented in Figure 10. show that atmospheric pressure at all altitudes from 0 to 15 km sharply increases in response to negative leap ΔE_{SW} and keeps this level in subsequent four days. In response to positive ΔE_{SW} the atmospheric pressure decreases at all altitudes at the zero day and then starts gradually to increase starting at 15 km at the first day and lowering to the ground level by the third day. In the context of the going down air masses flow it implies that the interplanetary electric field effectively accelerates or delays the descending tropospheric air masses.

To check this concept, the aerological measurements of the atmospheric winds at altitudes h = 0-25 km have been also analysed in [Troshichev, 2008]. The only horizontal wind parameters were measured at the air balloons launched at Vostok, and just these data were used to extract information about changes in the vertical flow of air masses. Figure 11. shows the mean vertical profiles of the horizontal wind mean direction and mean speed for "zero" day with large negative E_{SW} leap, for preceding -1st day (level of reference), and for succeeding 1st, 2nd, and 4th days. One can see that the wind azimuths lay in range 270 290° at h– 15-25 km, 240-270° at h=10-15 km, and 210-240° at h=5-10 km, i.e. the horizontal winds blow approximately toward the East.

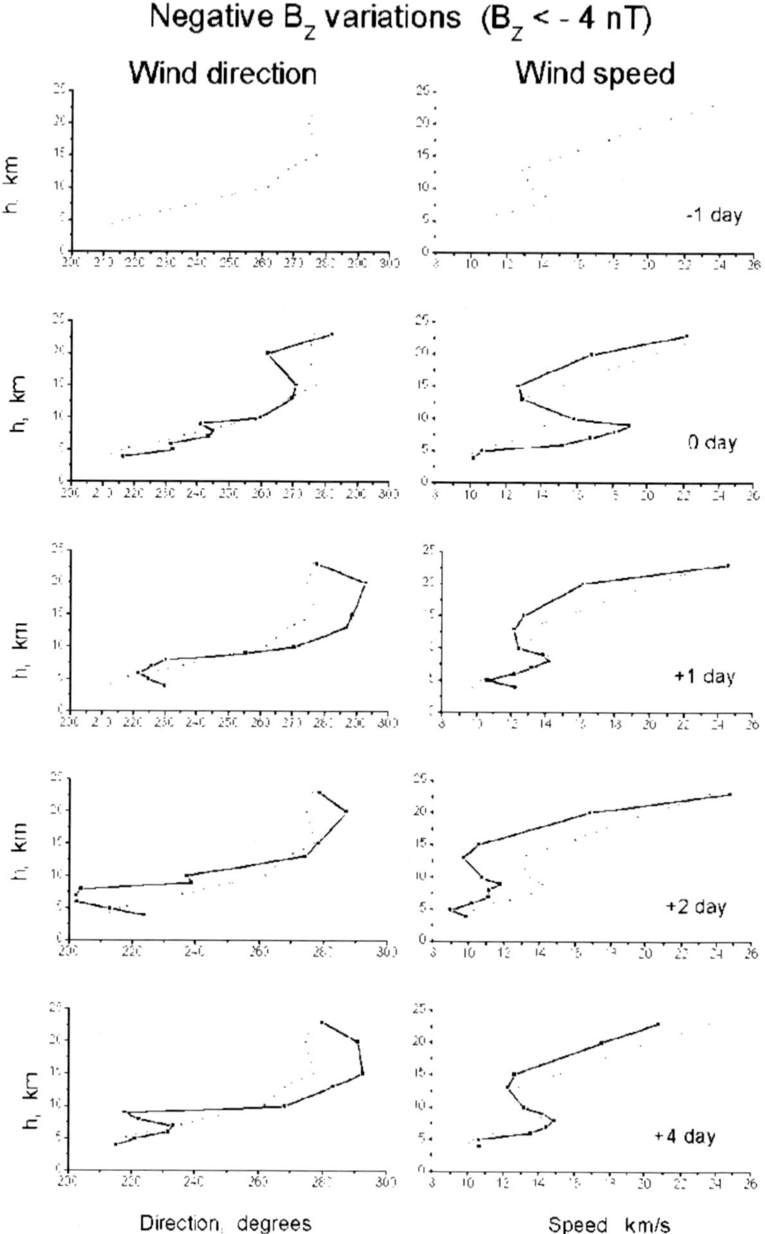

Figure 11. Dependence of the height profiles of the horizontal wind mean direction and mean speed above Vostok on the negative ΔE_{SW} influence. The wind profiles for the -1st day preceding the ΔE_{SW} impact (dotted lines) are taken as a level of reference [from Troshichev et al., 2004].

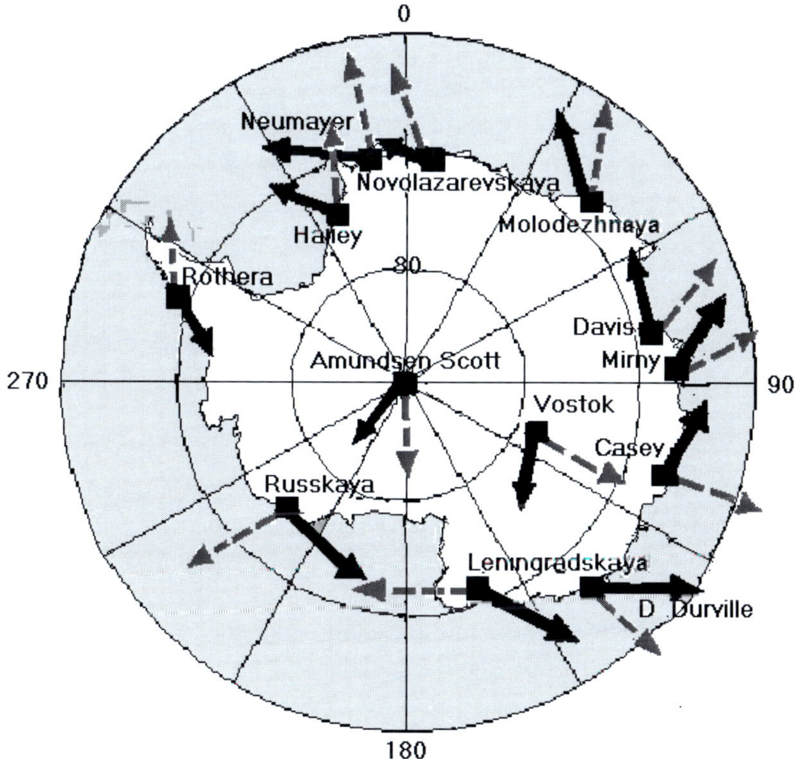

Figure 12. Distribution of the regular (solid arrows) and anomalous (dashed arrows) winds at the Antarctic stations [from Troshichev et al., 2008].

To understand this regularity we have to remember that the masses falling vertically in the rotating Earth's system are deviated toward the East under the action of Corioles force. Speed of this deviating motion is determined by formula $v_e = \omega \times r$, where ω is the angle speed of the Earth's rotation (ω = 1/240 °/s), and r is radius-vector in the point of observation (r ~ 880 km at latitude of Vostok station, φ = 82°). Taking into account these quantities and making allowance for angle ~ 8° between vectors ω and r, a rough measure of the deviation speed is obtained, $v_e \approx 10$ m/s. It is just mean value of the horizontal speed that was observed in aerological observations at altitudes 10-15 rm above Vostok station. Thus, we can suggest that direction and speed of the regular wind in the surface atmosphere at a top of the Antarctic ice dome is determined by deviating action of Corioles force on the descending air masses.

In response to the negative ΔE_{SW} leap the wind horizontal speed decreases in a key ("zero") day at altitudes higher 12 km and sharply increases at altitudes from 5 to 12 km (up to 35 % at h = 8-9 km), the vertical wind profile for -1st day being taken as a level of reference. The opposite regularity is typical of the positive ΔE_{SW} leaps. By the forth day the wind speed recovers to the quiet level. Taking into account that Corioles acceleration should be directly dependent on speed of the descending air masses, we suggest that increase or decrease of the wind horizontal speed is indicative of acceleration or slowing-down of the descending air masses. In such a case we came to conclusion that formation of cloud layer at h=5-10 km is favourable to acceleration of the descending masses, whereas clear sky favours their slowing-down. The mechanism of this linkage is unclear.

The acceleration (slowing-down) process should lead to the strong enhancement (or reduction) of the atmospheric pressure just for day of the negative (positive) ΔE_{SW} leap, as we have seen in Figure 10. It should be noted that the enhanced atmospheric pressure keeps during some days in the surface level (~ 3.5 km at Vostok station). One might expect that occurrence of the enhanced atmospheric pressure in the Central Antarctica will led to violation of the wind regime above the whole Antarctica.

Anomalous Winds at the Antarctic Stations and Their Relation to the IMF B_Z

The wind azimuth distribution at each Antarctic station is a specific feature, which is determined by the large-scale katabatic wind system and by local orography at a particular coast stations. Two separated extremes in the wind occurrence are observed at the coast stations during the winter season. The main noticeable extreme at 60-120° corresponds roughly to the westward wind formed due to action of the Corioles force on the moving radially drainage flow. The secondary minor extreme nearby 180° corresponds to the anomalous wind flowing from the Antarctic coast toward equator. One wide and rather flat maximum in a range from 180° to 270° is typical of the near-pole station Vostok, but the high-speed winds (V > 6 m/s) are observed exclusively at azimuth around 200°. In view of it, the winds at Vostok station falling within the range 180-210° were examined as anomalous winds. Figure 12. shows directions of the main and anomalous winds plotted on the map of Antarctica for all 13 stations. Thus, two wind patterns are typical of the winter

Antarctica: the main pattern is the regular circumpolar circle surrounding the continent, and the secondary pattern is system of the "anomalous winds" flowing from Antarctica toward the equator.

Analysis of relationship between the anomalous wind occurrence and changes in IMF B_Z component demonstrated [Troshichev et al., 2008] that the anomalous winds were preceded by the southward IMF coupling with the magnetosphere. Figure 13. shows, as an example, behavior of the mean anomalous wind observed at near-pole station Vostok in 1981-1989, and the mean anomalous winds observed simultaneously at the coast stations Neumayer, Casey, Russkaya, situated, correspondingly, in the Atlantic, Indian and Pacific ocean sectors.

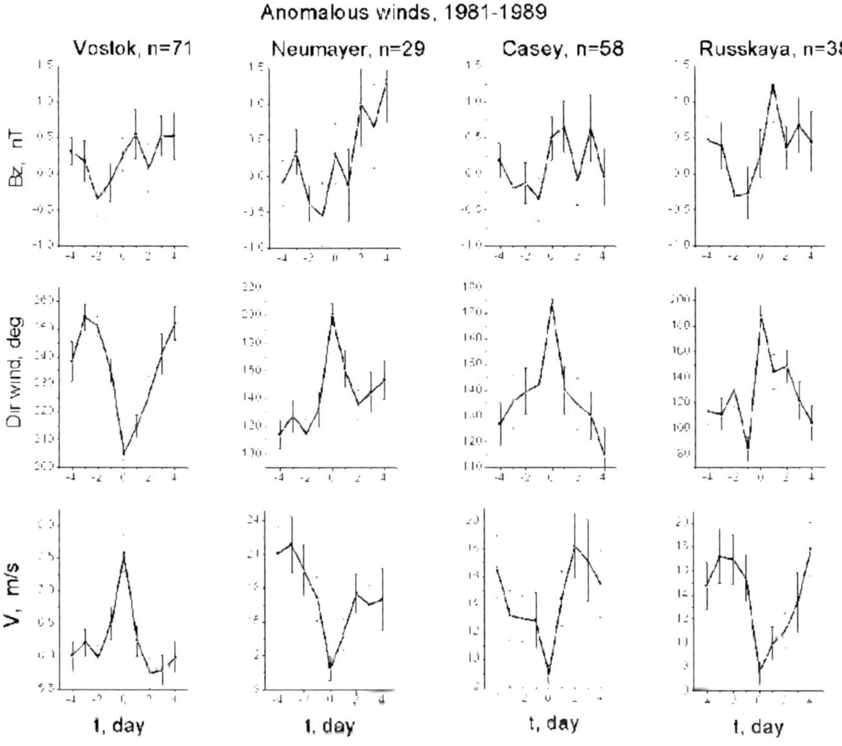

Figure 13. Anomalous winds at stations Vostok, Neumayer, Casey and Russkaya in their relation to changes in the IMF B_Z for winter seasons of 1981-1989, the day with anomalous wind occurrence at each particular station being taken as a key date [from Troshichev et al., 2008].

The mean variation of the appropriate IMF B_Z component was calculated separately for each station taking into account only the events with anomalous wind observed at this certain station. The day with anomalous wind occurrence at each particular station was taken as a zero date.

One can see from Figure 13. that occurrence of the anomalous winds at the coast stations was preceded by the increase of southward IMF B_Z and was delayed one-two days relative to the wind strengthening at Vostok. Not all wind violations starting at the Antarctic Ridge are extended to cover the overall coast area: only 82% of anomalous winds detected at Vostok reach Casey, located in the Indian ocean sector and as few as 41% reach Neumayer, located in the Atlantic sector.

It seems likely that extent of the coast station involvement in the process of the wind violation varies from one event to other. The anomalous winds at the coast stations, associated with high speed ($V > 6$m/s) anomalous wind at Vostok, were accompanied by the sharp reduction of the westward wind speed and by the succeeding wind turning toward the equator. This regularity corresponds to decay of the circumpolar vortex while increasing the katabatic winds, noticed in [Yasunari and Kodama, 1993]. The efficiency of the anomalous winds influence on the circumpolar vortex turned out to be dependent on duration of the southward IMF action: the 3-days influence of southward IMF with average $\Delta B_Z \sim 3$ nT (10 events) preceded the anomalous winds, which were observed simultaneously at the all coast stations.

SOI And Its Relation To Anomalous Wind System In Antarctica

Southern Oscillation (SO) is determined [Philander and Rasmussen, 1985] as a negative correlation between the sea level pressure fluctuations in the Southeast Pacific high and the North Australian-Indonesian low. A coupled system linking an anomalous warming of surface water in the eastern Pacific (El Niño) to an atmospheric branch SO, was named ENSO. During the years between warm events the opposite regularity often occurs and a cold phase of ENSO, the La Nina, exists [Van Loon and Shea, 1987]. Nature of the ENSO action is unknown. Stable links between Southern Oscillation and atmospheric processes in Antarctica was revealed in many studies [Trenberth, 1980; Mo et al., 1987; Van Loon and Shea, 1985, 1987; Parish and Bromwich, 1987; Bromwich et al., 1993; Smith and Stearns, 1993].

The conclusion was made, that propagation of the katabatic winds from Antarctica is a phenomenon that is accompanied by changes that involve the entire southern hemisphere [Bromwich et al., 1993]. To characterize the phase and intensity of the ENSO activity the Southern Oscillation Index (SOI) is used. SOI presents difference between the monthly pressure anomalies at Tahiti (Central Pacific) and Darwin (North Australia) [Van Loon and Shea, 1987]. The SOI is a nondimensional index, which is negative when ENSO is in a warm phase (El Niño events) and positive when ENSO is in a cold phase (La Nina events).

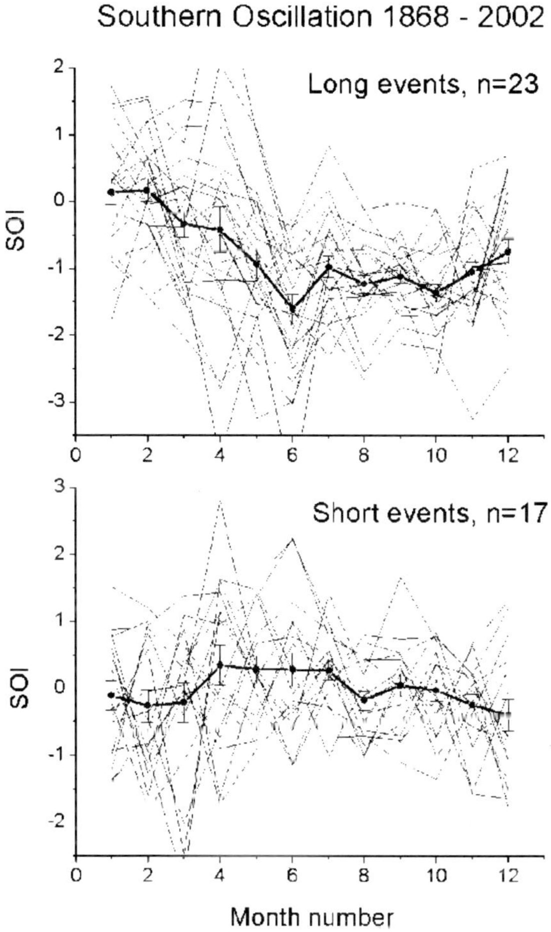

Figure 14. Behavior of SOI index in course of large negative long-lived and short-lived SOI deviations during 1868-2002 [from Troshichev et al., 2005].

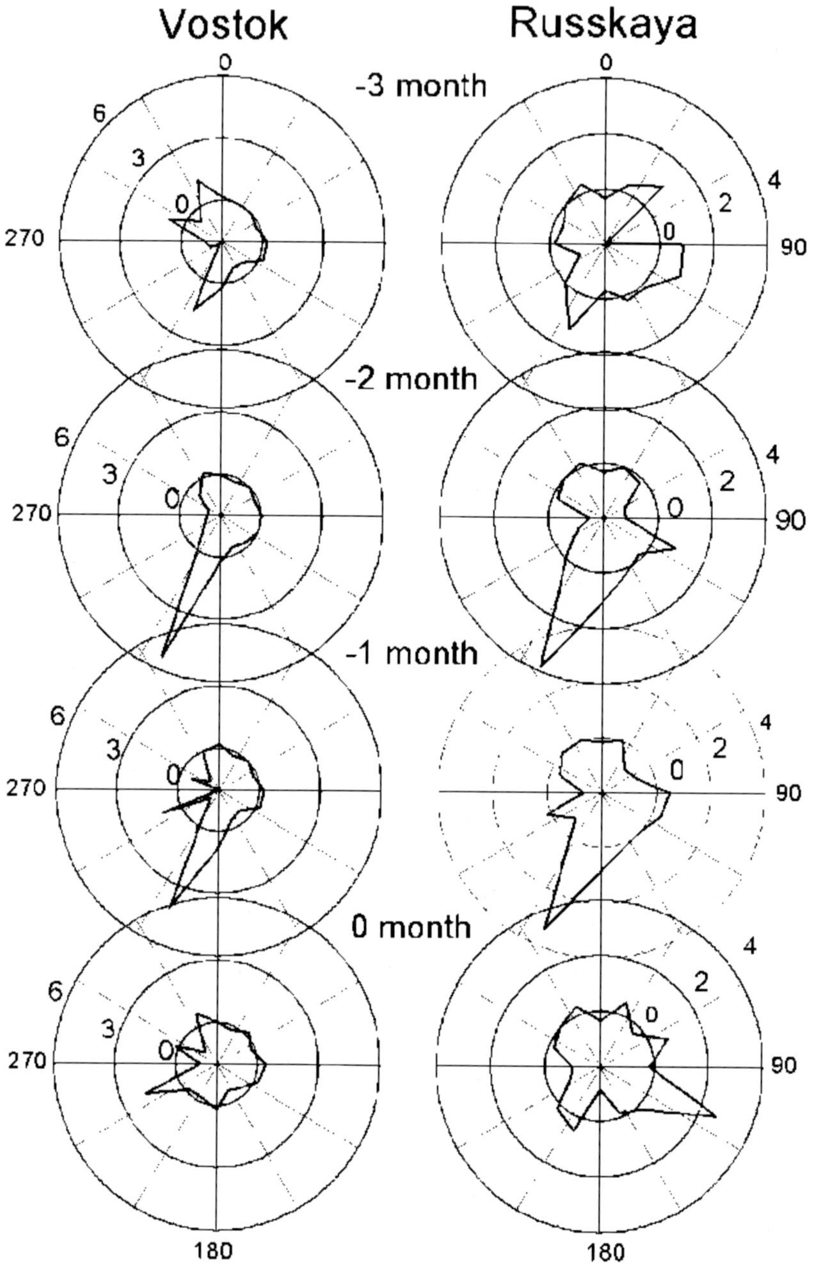

Figure 15. Differential wind roses, characterizing the changes in wind occurrence in course of El-Niño events for stations Vostok (a), and Russkaya (b) [from Troshichev et al., 2005].

One of the most intriguing feature of El Niño events is the seasonal regularity in their occurrence: formation of ENSO occurs mainly during the southern autumn-winter season [Van Loon and Shea, 1987]. Troshichev et al. [2005] examined all large negative SOI deviations during 1868-2002 and separated them in two groups in accordance with their duration: the short-lived deviations, lasting less than 3 months (17 events), and long-lived deviations, lasting more 6 months (23 events). It turned out that the long negative deviations (Figure 14), which correspond to real El-Nino events, start usually between March and June, and their mean early variation (thick line) reaches the minimum in June and tends to restore in November/December. The short deviations of SOI can begin in any time of year and their mean monthly values are close to zero.

To demonstrate the possible relation of SOI to the anomalous winds the mean angular distribution of the monthly winds ("the wind roses") at each station were examined for months preceding and succeeding the El-Niño beginning [Troshichev et al., 2005]. The El-Niño events with sharp onset (decrease of SOI value more than 1 during the month) have been only included in the examination. The wind rose for the –4th month, preceding the El-Niño onset, has been taken as a level of reference for all succeeding months, and the appropriate "differential wind roses", characterizing the changes in the wind azimuths for -3^{d}, -2^{d}, -1st, and zero (El-Niño onset) months have been constructed. Results of the analysis testified that the evident excess of winds above the level of reference was observed at angles 195-210° for Vostok, 185-215° for Russkaya, and 270-300° for Leningradskaya, which just correspond to azimuths of the anomalous winds at these stations. As Figure 15. shows, the anomalous winds are observed during -2^{d} and -1st months, preceding the El-Niño onset, and did not typical of -3th and zero months (as well as of succeeding months, which are not shown in Figure 15).

Availability of statistically significant relationships between the disturbed solar wind and anomalous winds above the Antarctica, on the one hand, and between the anomalous winds and SOI, on the other hand, makes it possible to suggest that linkage between the disturbed solar wind and development of SOI would be presented. Unfortunately, the monthly IMF data are very incomplete for years preceding 1998, and the relation of SOI to AE index (instead of IMF) was examined in [Troshichev et al., 2005]. Only sharp SOI events, in which the monthly value of SOI (negative or positive) was changed by 1 or more, have been included in the analyses. The month of this sharp change has been taken as a zero date in the superposition analysis.

Figure 16. shows behavior of SOI index for three types of changes in the southern oscillation:

a) Sharp declination of SOI with subsequent keeping of negative value during many months (basically, it is real El-Niño events with sudden onset) observed in 1969, 1972, 1982, 1986, 1991, 1994;
b) Sharp short-lived declination of SOI (1959, 1961, 1979, 1980, 1984, 1985); and
c) Short-lived increase of SOI (1962, 1964, 1970, 1975). The low panel in Figure 16. presents the proper changes in AE index. One can see, that in case of long negative SOI deviations the mean magnetic activity starts to increase 2-3 months before the beginning of El-Niño and reaches maximum just in month of the beginning (s.s. = 0.96).

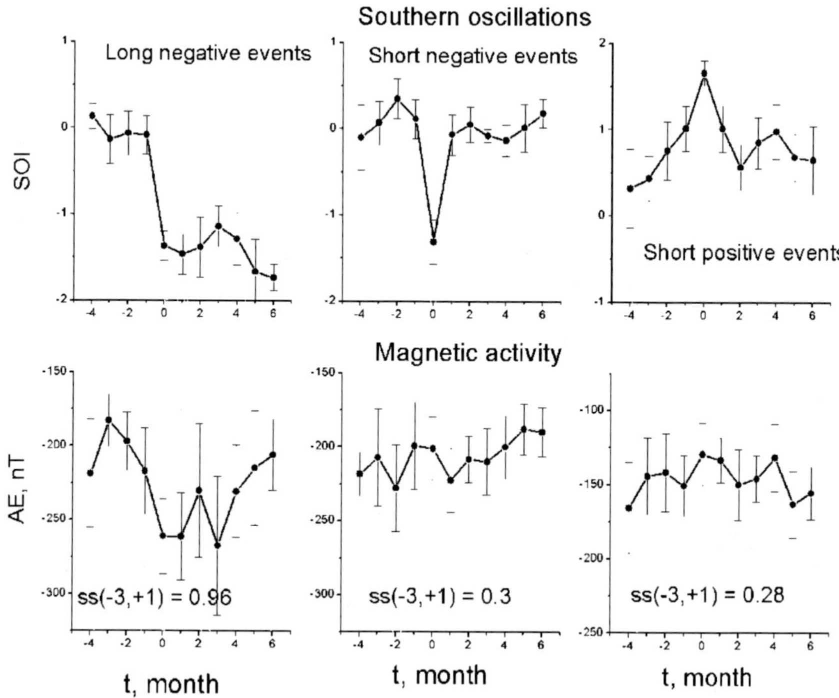

Figure 16. Relationship between changes in monthly values of SOI and AE indices for long-lived negative deviation of SOI (a), and short-lived negative and positive deviation of SOI (b) [from Troshichev et al., 2005].

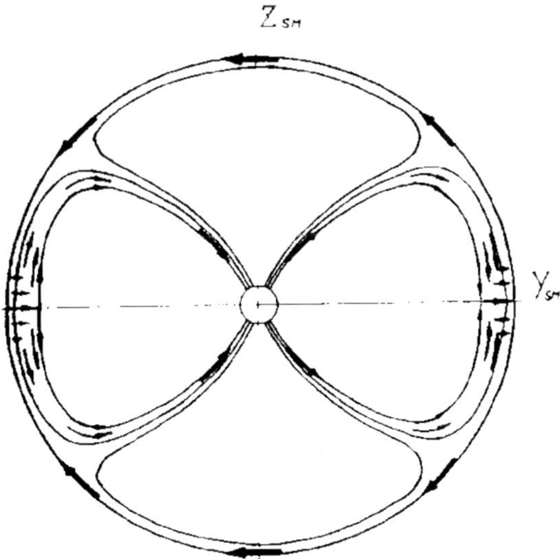

Figure 17. A conceptual sketch of generation of the electric currents in the low-latitude boundary layer, while the solar wind coupling with the magnetosphere [adopted from Troshichev, 1982]. It is shown the magnetosphere dawn-dusk cross viewed from the Sun. The arrows denote the electric currents. The field-aligned currents, flowing in the southern and northern polar caps at the dawn side and flowing out at the dusk side, provide the polar cap electric voltage.

On the contrary, in cases of short-term positive or negative deviations of SOI, the magnetic activity does not show the noticeable changes before and after the SOI impulses.

These results make it possible to conclude, that changes, negative or positive, in the southern oscillation occur irrespective of the solar wind influence, but development of the veritable El-Niño events, happening during southern winter, is likely influenced by the intense and lasting disturbances in the interplanetary electric field and correlate with corresponding magnetic activity.

Mechanisms Suggested to Explain the Solar Wind Influence on the Atmospheric Processes

A suggestion about the influence of the interplanetary electric field on cloudiness in Antarctica was made in [Troshichev and Janzhura, 2004] and based on the well-established fact of the solar wind impacting the magnetosphere. While the magnetosphere – solar wind coupling the interplanetary electric field, bearing by solar wind, generates the field-aligned currents, connecting the boundary magnetosphere with the polar ionosphere. The field-aligned currents flow into the polar ionosphere at the dawn side and flow out of the ionosphere at the dusk side, and provide the "overhead" polar cap ionospheric potential. The corresponding impact scheme is presented in Figure 17, adopted from [Troshichev, 1982]. Although some details of this process continue to be unknown, the linkage between the interplanetary electric field and the polar cap voltage is principally resolved and well defined.

Sharp changes of the ionospheric potential should intensify or reduce the electric currents between the polar ionosphere and the surface, passing through the layer at 5-8 km, where the atmospheric conductivity sharply declines [Handbook of Geophysics, 1985]. The connecting link between the polar cap voltage and the polar atmosphere is realized by the global electric circuit. There is a constant potential difference ~ 250 kV between the ionosphere and the Earth's surface, which is provided by the tropic thunderstorms. This potential drives the return downward currents (see Figure 18), which are most intense and variable in the polar areas (3-5 pA/m2) due to the effects of cosmic and magnetospheric energetic particles, and, what is more important, the influence of the polar cap voltage.

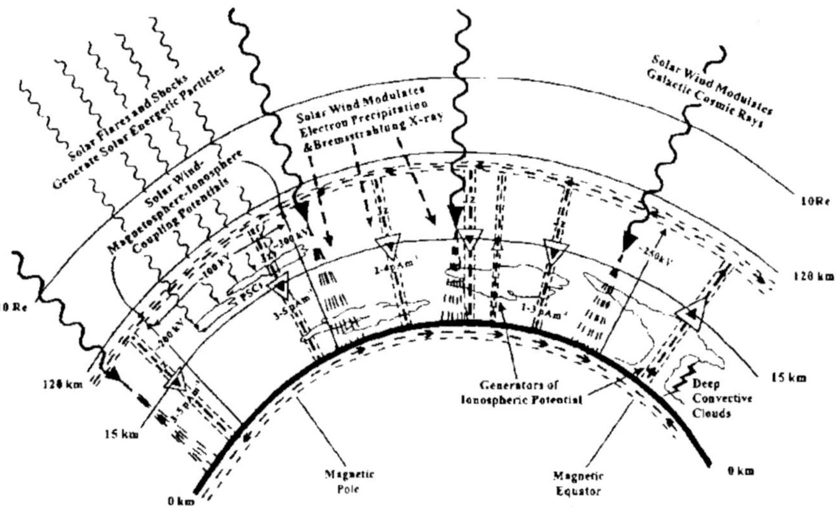

Figure 18. Diagram illustrating the global atmospheric electric circuit and the causes of its temporal and spatial variation (from [Tinsley and Zhou, 2006]). The vertical scale is greatly exaggerated below 120 km and greatly compressed above.

The influence of the solar wind on the global electric circuit is well documented [Tinsley, 1996]. The actual changes of the polar cap atmospheric electric field represent the combination of so-called Carnegie curve (describing the daily course of the tropic thunderstorms) and the deviations controlled by the E_{SW} changes [Frank-Kamenetsky et al., 1999, 2001]. It means that the interplanetary electric field affects the "overhead" ionospheric potential above the station.

According to Tinsley and Deen [1991] and Tinsley and Heelis [1993], who examined the cloud layer properties in relation to the global electric circuit, the space charges are accumulated at the boundary of sharp changes of atmospheric conductivity, in proportion to the downward current density through the cloud. It was shown that the ionosphere-Earth current density is modulated, particularly, by the polar cap ionospheric potential [Tinsley and Zhou, 2006]. The changes in tropospheric ionization might affect the rate of freezing of supercooled water droplets in high clouds. Since the mechanism of electrofreezing acts irrespective of the solar irradiation input, it would be workable also in the near pole region under conditions of the polar night.

CONCLUSION

Thus, the concept of the solar wind influence on processes in the Antarctic atmosphere seems to be convincingly verified with use of all available meteorological and aerological observations [Troshichev et al., 2003, 2004, 2005, 2008; Troshichev and Janzhura, 2004; Troshichev, 2008]. The disturbed solar wind has a greater impact on atmosphere processes in the central Antarctica, where the large-scale system of vertical circulation is formed during the winter seasons. The impact is realized through the interplanetary electric field E_{SW}, which influences the magnetosphere field-aligned currents, creating the appropriate polar cap electric voltage. The ionospheric electric potential strongly affects the global electric circuit, which is generated by the tropic thunderstorms and is closed mainly in polar caps.

Two processes are brought into operation when the IMF B_{ZS} component and the appropriate electric field E_{SW} strongly increase. The first of them is formation of the cloud layer above the Antarctic Ridge, where the air masses income into the central Antarctica from above troposphere. Under the usual conditions the atmosphere on the Antarctic ridge is in state of the thermal quasi-equilibrium owing to superposition of the adiabatic warming of the descending air masses and stable radiation cooling of air situated at the ice sheet. The cloud layer will efficiently backscatter the long wavelength radiation going from the ice sheet, but it will not affect the adiabatic warming process. As a result of the radiative cooling reduction, the atmosphere should be warm below the cloud layer and would be cool above the layer. Just such regularity is observed during specific events of the sudden warmings, happening from time to time in the central Antarctica.

The second process is acceleration of the air masses incoming into the central Antarctica from troposphere. This process causes the sharp increase of the atmospheric pressure in the surface layer and gives rise to collapse of the regular large-scale wind system above the entire Antarctica. The draining winds are strongly strengthened and the circumpolar vortex about the periphery of the Antarctic continent correspondingly decays.

Both processes are strongly linked and lead to the cold air masses flow out to the Southern ocean. The latter process evidently destroys the regular relationships between the sea level pressure fluctuations in the Southeast Pacific high and the North Australian-Indonesian low and promote the El-Niño beginning.

References

Bromwich, D. H., J. F. Carrasco, Z. Liu and R. Y. Tzeng: Hemispheric atmospheric variations and oceanographic impacts associated with katabatic surges across the Ross Shelf, Antarctica. *J. Geophys. Res.*, 98(D7), 13045-13062 (1993).

Egger, J.: Slope winds and the axisymmetric circulation over Antarctica. *J. Atmos. Sci.*, 42, 1859-1867 (1985).

Farrar, P. D.: Are cosmic rays influencing ocean cloud coverage – or is it only El Nino? *Climate Change.*, 47, 7-15 (2000).

Frank-Kamenetsky, A. V., G. B. Burns, O. A. Troshichev, V. O. Papitashvili, E. A. Bering and W. J. R. French: The geoelectric field at Vostok, Antarctica: its relation to the interplanetary magnetic field and the cross polar cap potential difference. *J. Atmos. Solar-Terr. Phys.*, 61, 1348-1356 (1999).

Frank-Kamenetsky, A. V., O. A. Troshichev, G. B. Burns and V. O. Papitashvili: Variations of the atmospheric electric field in the near-pole region related to the interplanetary magnetic field. *J. Geophys. Res.*, 106, 179-190 (2001).

Handbook on Geophysics and the Space Environment, (Eds Jursa, A. S.). Air Force Geophysical Laboratory, U. S. A. F. (1985).

Harrison, R. G. and K. S. Carslaw: Ion-aerosol-cloud processes in the lower atmosphere, *Rev. Geophys.*, 41, 1012-1026. doi: 10.1029/2002RG000114 (2003).

Herman, J. R. and R. A. Goldberg: Sun, Weather, and Climate. N. A. S. A., Washington, D. C., 430 p. (1978).

James, I. N.: The Antarctic drainage flow: Implications for hemispheric flow on the Southern Hemisphere. *Antarct. Sci.*, 1, 279-290 (1989).

Kristjansson, J. E., A. Staple, J. Kristiansen and E. Kaas: A new look at possible connection between solar activity, clouds and climate, *Geophys. Res. Let.*, 29(23), 2107, doi:10.1029/2002GL015646 (2002).

Marsh, N. and H. Svensmark: Galactic cosmic ray and El Nino-Southern Oscillation trends in International Satellite Cloud Climatology Project D2 low-cloud properties, *J. Geophys. Res.*, 108(D6), 4195 doi: 10.1029/2001JD 001264 (2003).

Mo, K. C., J. Pfaendtner and E. Kalnay: A G. C. M. study on the maintenance of the June 1982 blocking in the southern hemisphere. *J. Atmos. Sci.*, 44, 1123-1142 (1987).

Palle, E. and C. J. Butler: The proposed connection between clouds and cosmic rays: cloud behavior during the past 50-120 years. *J. Atmos. Solar-Terr. Phys.*, 64(3), 327-337 (2002).

Parish, T. R. and D. H. Bromwich: The surface windfield over the Antarctic ice sheets: *Nat.*, 328, 51-54 (1987).

Parish, T. R. and D. H. Bromvich: Continental-scale simulation of the Antarctic katabatic wind regime. *J. Climate.*, 4, 135-146 (1991).

Parish, T. R.: On the role of Antarctic katabatic winds in forcing large-scale tropospheric motions. *J. Atmos. Sci.*, 49, 1374-1385 (1992).

Philander, S. G. and E. M. Rasmusson: The southern oscillation and El-Nino, *Adv. Geophys.*, 28(A), 197-215 (1985).

Pudovkin, M. I. and S. V. Veretenenko: Cloudness decreases associated with Forbush-decreases of the galactic cosmic rays. *J. Atmos. Terr. Phys.*, 57, 1349-1355 (1995).

Pudovkin, M. I., S. V. Veretenenko, R. Pellinen and E. Kyro: Cosmic ray variation effects in the temperature of the high-latitude atmosphere. *Adv. Space Research.*, 17(11), 165-168 (1996).

Pudovkin, M. I., S. V. Veretenenko, R. Pellinen and E. Kyro: Meteorological characteristic changes in the high latitudinal atmosphere associated with

Forbush decreases of the galactic cosmic rays. *Adv. Space Res.,* 20(6), 1169-1177 (1997).

Schwerdtfeger, W.: *Weather and Climate of the Antarctic*, 261pp., Elsevier, New York (1984).

Smith, S. R. and C. R. Stearns: Antarctic pressure and temperature anomalies surrounding the minimum in the southern oscillation index. *J. Geophys. Res.,* 98, 13071-13083 (1993).

Svensmark, H. and E. Friis-Christensen: Variation of cosmic ray flux and global cloud coverage - a missing link in solar climate relations, *J. Solar-Terr. Phys*., 59, 1225-1232 (1997).

Tinsley, B. A.: Correlations of atmospheric dynamics with solar wind induced changes of air-Earth current density into cloud tops. *J. Geophys. Res.*, 101, 29701-29714 (1996).

Tinsley, B. A., G. M. Brown, and P. H. Scherrer: Solar variability influences onweather and climate: possible connection through cosmic ray fluxes and storm intensification. *J. Geophys. Res*., 94, 14783-14792 (1989).

Tinsley, B. A. and C. W. Deen: Apparent tropospheric response to MeV-GeV particle flux variations: a connection via eletrofreezing of supercooled water in high-level clouds? *J. Geophys. Res.,* 96, 22283-22296 (1991).

Tinsley, B. A. and R. A. Heelis: Correlations of atmospheric dynamics with solar activity: evidence for a connection via the solar wind, atmospheric electricity, and cloud microphysics. *J. Geoph. Res.,* 98, 10375-10384 (1993).

Tinsley, B. A. and L. Zhou: Initial results of a global circuit model with variable stratospheric and tropospheric aerosols, *J. Geophys. Res.*, 111, 16205-16223 doi: 10.1029/2005JD006988 (2006).

Todd, M. and D. Kniveton: Changes in cloud cover associated with Forbush decreases of galactic cosmic rays, *J. Geophys. Res.,* 106, 32031-32041 (2001).

Trenberth, K. E.: Planetary waves at 500 mb in the southern hemisphere. *Mon. Weather Rev*., 108, 1378-1389 (1980).

Troshichev, O. A.: Polar magnetic disturbances and field-aligned currents. *Space Sci. Reviews*., 32, 275-360 (1982).

Troshichev, O. A., L. V. Egorova and V. Ya. Vovk: Evidence for influence of the solar wind variations on atmospheric temperature in the southern polar region. *J. Atmos. Solar-Terr. Phys*., 65, 947-956 (2003).

Troshichev O. A., L. V. Egorova and V. Ya. Vovk: Influence of the solar wind variations on atmospheric parameters in the southern polar region. *Adv. Space Res*., 34, 1824-1829 (2004).

Troshichev, O. and A. Janzhura: Temperature alterations on the Antarctic Ice sheet initiated by the disturbed solar wind. *J. Atmos. Solar-Terr. Phys*., 66, 1159-1172 (2004).

Troshichev, O. A., L. V. Egorova and V. Ya. Vovk: Influence of the disturbed solar wind on atmospheric processes in Antarctica and El-Nino Southern Oscillation. *Mem. Soc. Astronomy of Itali*a., 76, 890-898 (2005).

Troshichev, O., V. Vovk and L. Egorova: I. M. F. associated cloudiness above near-pole station Vostok: impact on wind regime in winter Antarctica, *J. Atmos. Solar Terr. Phys*., 70, 1289-1300 (2008).

Van Loon, H. and D. J. Shea: The Southern Oscillation, IV: The precursors south of 15°S to the extremesof the oscillation, Mon. *Weather Rev.,* 113, 2063-2074 (1985).

Van Loon, H. and D. J. Shea: The Southern Oscillation, VI, Anomalies of sea level pressure on the southern hemisphere and of Pacific sea surface temperature during the development of a warm event, Mon. *Weather Rev.,* 115, 370-379 (1987).

Wilcox, J. M.: Solar activity and weather, *J. Atmos. Terr. Phys*., 37, 237-243 (1975).

Yasunari, T., and S. Kodama: Intraseasonal variation of katabatic wind over East Antarctica and planetary flow regime in the southern hemisphere. *J. Geophys. Res*., 98, 13063-13070 (1993).

In: Solar Wind
ISBN 978-1-62081-979-1
Editors: C. D. E. Borrega, A. F. B. Cruz

Chapter 5

EXPERIMENTAL AND MODELING EVIDENCES OF THE SOLAR WIND ENERGY INFLUENCE ON THE EARTH ATMOSPHERE

L. N. Makarova[*] *and A. V. Shirochkov*
Arctic and Antarctic Research Institute, St.Petersburg, Russian Federation

ABSTRACT

So far the solar wind energy contribution to energetic balance of the Earth atmosphere was ignored in any atmospheric and climatic research. However the solar wind is a permanent source of a significant amount of the electromagnetic energy emitted by the Sun which is constantly supplied to the near-Earth space. Traditionally this energy was attributed entirely to sustain a definite level of geomagnetic activity expressed as intensity of the geomagnetic substorms and storms. The authors of this paper found in 1997 after analysis of the data of the Russian rocket sounding in the Arctic that enhancement of the solar wind dynamic pressure do influence thermal regime of the polar middle atmosphere. Similar analysis of the atmospheric balloon sounding data obtained at different stations in both the Arctic and the Antarctica shows that the stratospheric temperature closely correlated with the solar wind electromagnetic energy. After establishing these statistically confident

[*] Correspondence to: Arctic and Antarctic Research Institute, St.-Petersburg, 199397 Russia, phone 7-812-352-06-01, fax 7-812-352-26-88 e-mail: *shirmak@aari.nw.ru.*

relations it was necessary to find a plausibly reasonable physical mechanism which could explain reality of the found coupling. A concept of the global electric circuit as a physical mechanism for explanation of a direct coupling between the solar wind and the middle atmosphere was suggested. We proposed a new, modified version of the global electric circuit with two Electro-Motive Force (EMF) generators: internal EMF generator driven by the thunderstorm activity of the Earth (a common feature of previous circuit configurations) and an external EMF generator driven by the solar wind energy. The passive elements of this circuit are the ionospheric E-layer (external element of previous version of the circuit), stratospheric conducting layer of heavy ions (h=20-25 km) and conducting layer of the Earth surface. In this configuration a previous scheme of the global electric circuit is a part of the proposed version of it. Numerical evaluation of the electromagnetic energy of the solar wind is a very difficult task. It can be done only approximately. Structure of the Earth magnetosphere is changing constantly upon influence of the solar wind as well as a position of a boundary of the magnetosphere (magnetopause). The problem could be solved if we present boundary of the Earth magnetosphere and the ground surface as a giant capacitor with external and internal plates correspondingly. The external plate of this capacitor (magnetopause) could be moved toward the Earth under the solar wind pressure. The energy of the solar wind roughly can be calculated by estimation of energy which is required to move the magnetopause for a definite distance. The magnetopause is located at ~ 12 Re (were Re is the Earth radius) under a quite condition of the solar wind. During strong disturbances of the solar wind the magnetopause could approach the Earth at distance of approximately ~ 6 Re. Such estimation shows that energy required for movement of magnetopause at a distance of 6 Re is equal to ~ 5 10 15J. Preliminary numerical estimations showed that under typical conditions such amount of the Joule heating dissipated in stratosphere is comparable with a rate of heating of ozone layer by the solar UV radiation. Furthermore, such amount of energy is sufficient for enhancement of cyclonic activity in the Earth atmosphere. As the next step of exploration a numerical calculation scheme was elaborated, which took into account the abovementioned processes. This numerical scheme was successfully used in one of the global dynamical photo-chemical models of the atmospheric circulations. The results of these model simulations confirmed all previously made preliminary estimations concerning influence of the solar wind energy on the atmospheric processes. There are the definite plans to improve the effectiveness of the proposed physical mechanism describing interaction of the solar wind with the Earth atmosphere. Evaluation of the effects of different degree of the Earth electric conductivity must be taken into account in the next explorations on the subject.

INTRODUCTION

So far the solar wind energy contribution to energetic balance of the Earth atmosphere was actually ignored in any atmospheric and climatic research. However the solar wind is a permanent source of a significant amount of electromagnetic energy emitted by the Sun which is constantly supplied to the near-Earth space. Traditionally this energy was attributed entirely to sustain a definite level of geomagnetic activity expressed as intensity of the geomagnetic substorms and storms. The most of previous studies of the solar-terrestrial physics were based on search of statistical connections between a level of solar activity expressed as the sunspot number and various indices of geomagnetic activity. Such attempts were met with a definite skepticism although sometimes obtained statistical relations happened to be rather positive. Basically any index of geomagnetic activity in accordance with well-known Biot - Savart physical law is related to intensity of ionospheric currents flowing in the E-region of the Earth's ionosphere (altitudes 100-120 km). However, no reliable physical mechanism which could explain declared statistical relations between solar activity and the Earth climatic processes has been proposed so far. A new era in studies of the solar-terrestrial relations began with start of regular satellite observations of the near-Earth space. The Space originated energy is universally adopted as a factor capable to control the Earth's climate dynamics. A level of the Sun UV radiation expressed as the total solar irradiance (TSI) or the "solar constant" is taken as the most reliable indicator of the amount of the solar energy transferred to Earth. A great advantage of this parameter as compared to traditional index of the solar activity level –Sun spot number (SSN) - is that it is based on direct instrumental observations of the solar irradiance intensity in a wide frequency interval above the Earth's atmosphere. Hence, the possibilities appeared to study the solar-terrestrial relations in both long-term and short-term intervals by using reliable parameter of TSI expressed in the direct energy units. On the other hand the Sun variability includes other electromagnetic emissions of different intensity and duration, which certainly contribute to the total energy of the solar wind - a well-established permanent component of the solar activity whose influence on the climate dynamics has been underrated so far. The solar wind permanently affecting near-Earth space could provide substantial amount of energy to sustain active atmospheric processes [6]. Quantitatively this energy could be evaluated crudely as the dynamic pressure of the solar wind. More accurately influence of the solar wind on the Earth's magnetosphere-ionosphere-atmosphere system could be expressed by means of a newly introduced parameter-subsolar distance between the Earth and external boundary of the terrestrial magnetosphere –magnetopause [12]. The latter authors presented evidence of the close connections between various climatic characteristics and the solar wind disturbances expressed as the solar wind dynamic pressure enhancements [10].

They showed that temperature and baric regimes of the high-latitude middle atmosphere as measured by rocket sounding at the Heiss Island (Franz-Josef Archipelago in the Arctic) have changed under various magnitudes of the solar wind dynamic pressure. It was probably the very first experimental evidence of direct influence of the solar wind disturbances on the middle atmosphere behavior. These authors claimed that these couplings could be explained within the framework of the revised model of the global electric circuit [4] with external electro-mobile force (EMF) generator, which is located at the dayside magnetopause and is controlled by the solar wind energy. More lately some indirect evidences of the solar wind influence on the atmospheric processes in the southern polar region were given in [18]. However no indication of any possible physical mechanism capable to explain these coupling was introduced in this paper.

It is clear that all these suggestions and conclusions need more exploration. The purpose of this paper is to demonstrate that the idea of a direct influence of the solar wind energy on the atmospheric processes is a plausible and reasonable suggestion. Several experimental and modeling evidences in favor of this statement will be presented together with indication of further steps in developing this idea.

Observations

It was shown previously that the solar wind energy evaluated as the subsolar distance between the Earth and magnetopause (expressed in the Earth's radius units) is connected closely with the thermal regime of stratosphere at the polar and middle latitudes [12]. Although these results were based on statistical studies, the authors proposed a reasonable physical mechanism capable to explain such connections. Nevertheless further experimental data are needed to prove reality of *the data presented on* Figure 1.

Figure 1 shows the results of the atmosphere balloon sounding made in Murmansk (68°.25N; 33°.48 E; invariant latitude 64°.35 N) during October 10, 1988 – December 25, 1988.

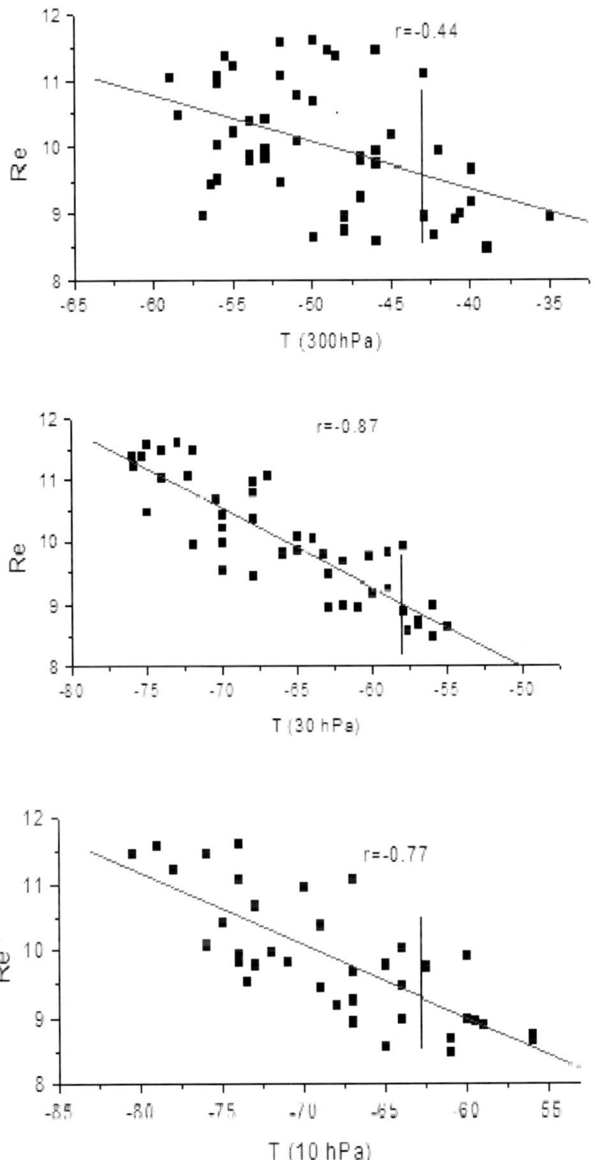

Figure 1. Variations of middle atmosphere temperature at Murmansk station in dependence on position of magnetopause relative to the Earth expressed in the Earth radius units Re.

The data of temperature measurements at baric surfaces 300 hPa (h=9 km), 30 hPa (h=23 km) and 10 hPa (h=28 km) made at local noon in winter are compared with the corresponding magnetopause position given in the Earth's radius units and calculated from a model presented in [16]. It is clear that stratosphere becomes warmer with the magnetopause approaching the Earth at all heights shown here. However the best correlation (r=0,87) between two data sets is observed at height of 23 km. Statistical significance of these relations was checked by means of the Student t-distribution test. A confidence interval for 95 % confidence level is shown on figure 1 by thin vertical *bars.* Above and below this altitude there is a notable decrease in degree of this correlation. It seems that this systematic change of the degree of correlation between the solar wind energy and stratospheric temperature has a solid physical foundation. The main feature of the version of the global electric circuit developed by [10, 12] is existence of the conducting layer in the middle atmosphere at the height of about 23 km. There are reliable experimental evidences (rocket sounding data) of the existence of such conducting layer formed by the heavy ion of cluster type [5, 15, 19]. Interaction of the solar wind energy with atmosphere in this scheme is as follows: under enhanced pressure of the solar wind during close approach of magnetopause to the Earth, excessive electric fields are generated. These fields induce electric current in the above-mentioned conducting layer of the middle atmosphere as a part of the global electric circuit. The Joule heating produced by this current contributes to the warming of the stratosphere during disturbances in the solar wind. These effects are concentrated in the core of the layer with reasonable diminishing of them at both higher and lower altitudes. The authors made numerical evaluations of these effects which will be presented in the following parts of the Chapter. However it is worth mentioning that these calculations supported explanations described above. Since the solar wind energy is proved to be geoeffective force, it is worth trying to determine its relation to other kinds of the solar activity whose influence on the Earth weather and climate is commonly accepted.

Parameterization of the Heating in the Middle Stratosphere Due to the Solar Wind Induced Electric Currents

Most studies of the solar effects on the Earth's atmosphere and climate paid attention to the radiative energy variations connected with 27-days and 11-year solar cycles [2]. The effects of the energetic particle precipitation induced by the solar wind disturbances on the middle atmosphere composition and dynamics have been considered also as a part of the solar output [6]. These processes can affect the ozone and other radiatively active atmospheric species and can be detected mostly in the upper and middle stratosphere. Analysis of the experimental data as well as theoretical considerations revealed that the solar wind energy could be transferred into the Earth's atmosphere also by the electric fields [6, 12].

A high correlation between the solar wind activity and different atmospheric parameters suggests a strong coupling of these parts of the near-Earth space [10, 11, 12]. A new view on the global electric circuit proposed by Makarova et al. [11, 12] could help to explain this close coupling. According to this approach the main sources of energy controlling the global electric circuit are two Electro-Motive Force generators driven by the thunderstorm activity of the Earth and by the solar wind energy correspondingly. Variations of parameters of the atmosphere are connected with changes of these two sources of electric energy with time. The passive elements of the circuit are the conducting regions in the ionosphere, a layer of heavy ion-clusters in the middle stratosphere and conducting parts of the ground surface.

Demonstration of reality of a physical mechanism determining influence of the global electric circuit on the atmospheric parameters is connected with solving the following problems:

1. Existence of a conducting ion layer in the stratosphere.
2. Quantitative estimation of the effects caused by the electric currents in the middle stratosphere.
3. Quantitative calculation of changes in thermal regime of the stratosphere caused by electric currents with the realistic input parameters.
4. Influence of conductivity of the ground surface as one of the main parameters of the global electric circuit.
5. Every part of these questions will be considered below.

1. Peak of Heavy Ion Density in the Earth's Stratosphere and its Influence on the Middle Atmosphere Processes

The first notes of reality of existence of the stratospheric heavy ion layer (so called cosmic layer C) appeared in scientific literature long time ago. This problem was explored theoretically as well as by means of special rocket mass spectrometer measurements in the middle atmosphere. For example, in paper [19] it was showed averaged altitude profile of electron concentration in the whole atmosphere with indication of the mean ions at each height. A definite maximum of the charged particles (heavy ions) at the heights around 30 km equal to ~ 5 x 10^3 cm^{-3} is clearly seen at these data. Brasseur and Chatel [5] demonstrated similar data but additional important information on the mean ion mass numbers is shown here. They showed that the heavy ion – clusters with ion mass numbers equal to 100 – 500 are the dominant charged particles at stratospheric heights (20 – 30 km).

Unfortunately this important information was obtained from the episodic rocket measurements made at the moderate latitudes of the Northern hemisphere. Rosen and Hoffmann [15] presented more long–term data concerning existence of stratospheric layer of charged particles.

These authors analyzed data of the balloon measurements of the ion content of the atmosphere by means of the Gerden collector method performed in two places: Saskatoon in Canada (geographic latitude 52° N) and Larami in Wyoming, USA (geographic latitude 41° N).

Maximum of ion concentration equal to N_i = ~5.5 x 10^3 cm^{-3} was steadily recorded at the height of~15 km. It is worthy to note that their data demonstrated almost unchanged shape of the profiles obtained during rather long period of measurements (December 1986–March 1987). Additional important feature of their measurements was disclosed in a case of simultaneous balloon measurements of height distribution of the positive and negative ions as well as ozone density at Larami on February 9, 1987 (their Figure 5). It was evident that concentration of the ions of both polarities are similar and maximum of their density is located at ~ 15 km. Maximum of the main ozone layer is located by several kilometers above the ion maximum. It seems that a close neighborhood of these two important atmospheric parameters is not a casual phenomenon. So, it is possible to claim that the stratospheric ion distribution with maximum equal to N_i =~5.5 x 10^3 cm^{-3} located at the heights 15 – 20 km is a permanent feature of the middle atmosphere.

There are two aspects of the problem of interaction of cosmic rays with stratosphere:

a) Production of permanent maximum in stratospheric ion distribution, and
b) Peculiarities of the specific physical and chemical processes in stratosphere under intense impact of the cosmic rays.

The former problem could be solved theoretically if one knows values of density of charged particles produced by intruding fluxes of cosmic rays. We used for this purpose magnitudes of the cosmic ray fluxes given by Bazilevskaya et al. [3] as well as values of the rates of ionization produced by these fluxes given in the same paper. The density of the charged particles in stratosphere was calculated by using equation of ion – recombination equilibrium:

$$\alpha_{eff} = q/ N_i^2 \quad (1)$$

where: α_{eff} – effective recombination coefficient;

q – rate of ionization,

N_i – density of the charged particles.

The values of α_{eff} were taken from Rosen and Hofmann [15]. We calculated height profiles of the charged particles in the stratosphere by using this equation.(They are not shown here). The profiles of N_i at the high geomagnetic latitudes (λ > 65°) demonstrate two maximums: the first one, equal to ~7 x10^3 cm^{-3} and located at h = 30 km and the second one, equal to 6 10^3 cm^{-3} and located at h ~ 15 km. Concentration of the charged particles in the stratosphere decreases with decrease of geomagnetic latitude due to geomagnetic field influence. It is worthy to note that calculated value of N_i for moderate latitude is almost identical to experimental data obtained by Rosen and Hofmann [15] for Saskatoon and Larami.

So, we proved theoretically existence of maximum in stratospheric ion distribution by using independent reliable experimental data of the cosmic ray fluxes measured by atmospheric balloons at different geomagnetic latitudes.

It is important to note that interaction of the neutral and charged particles in stratosphere occur under condition of very high density (10^{17}– 10^{18} cm^{-3}) of the neutral atmospheric components.

Therefore the main attention must be given to processes of collision between neutral and charged particles. Elementary kinetic theory [21] determines a value of mean free path of particles λ_a with radius r_a and distance between the centers of the particles $2r_a$ by the following equation:

$$\lambda_a = 1/\pi\, (2r_a)^2\, N_a \quad (2)$$

where: N_a – concentration of neutral molecules.

If electron with very small radius moves amidst neutral molecules of gas we can assume that collisions occur when distance between electron and neutral molecule is equal to r_a. In this situation electron changes direction of its movement and loses its energy. Therefore mean free path of electron under the circumstances could be expressed as:

$$\lambda_e = 1/\pi (r_a)^2 N_a \quad (3)$$

Small radius of the electron allows one to ignore collisions of electrons with each other. We will obtain equation expressing magnitude of collision "electron – neutral" ν_{ea}:

$$\nu_{ea} = V_e/\lambda_e = \pi (r_a)^2 N_a (2kT/m)^{1/2} \quad (4)$$

if we take into account dependence of electron velocity V_e on temperature equal to:

$$V_e = (2kT/m)^{1/2} \quad (5)$$

This equation (4) shows that value of collision "electron – neutral" depends on square of radius of neutral component. It becomes to be very important for stratosphere with its giant molecules clusters of hydroxyls which size could be as great as 100 Å, while usual atmospheric molecules size is ~ 1 – 3 Å (10^{-8}). It is reasonable to assume that stratospheric ions could also influence on the electron trajectory by means of collision "ion – electron".

Radius of ion impact is greater than of neutral particles r_a due to insignificant diminishing of Coulomb forces with distance.

It is reasonable to assume that electron will collide with ion at some distance r_{eff} when its kinetic energy will be equal to potential energy, i.e.

$$3/2kT \sim e^2/\varepsilon_0 r_{eff} \quad (6)$$

where T - temperature of atmosphere;

k – Boltzmann constant;

e - charge of electron;

ε_0 – dielectric constant.

Cross – section of collision "electron – ion" determined by magnitude of r_{eff} depends on electron velocity. Taking into account these facts one can calculate numerically value of collision frequency "electron – ion" by means of equation:

$$\nu_{ei} = \pi\,(2e^2/3\varepsilon_0\,kT)^2\,N_i\,(3kT/m)^{1/2} \quad (7)$$

The total value of collision frequency of electron with both neutral particles and ion could be expressed as:

$$\nu_e = \nu_{ea} + \nu_{ei} = \pi\,(r_a)^2\,N_a\,(2kT/m)^{1/2} + \pi\,(2e^2/3\varepsilon_0\,kT)^2\,N_i\,(3kT/m)^{1/2} \quad (8)$$

If we take $r_a = 10^{-8}$ cm, $N_a = 10^{18}$ cm^{-3}, $N_i = 5 \times 10^3$ cm^{-3}, T = 220°K and substitute them into correspondent expression for collisions we will obtain:

$$\nu_{ea} = 1.8 \times 10^{-8}\,N_a\,(T/300) \sim 3 \times 10^9\,s^{-1}$$

$$\nu_{ei} \cdot 1 \times 10^{-2}\,N_i\,(300/T)^{3/2} \sim 10^2\,s^{-1}$$

So the time between collisions of electrons with neutral particles is ~ 10^{-9} s while the corresponding period between collisions of electrons with ions is ~ 10^{-2} s. It means that stratospheric electrons with additional energy obtained from electric field transform this energy to neutral particles for period of time, which is much shorter than photochemical interaction of charged particles with neutral components (this period is equal to ~ 10^{-3} s). This fact is a strong confirmation of reality of physical mechanism of transforming of the solar wind energy into middle atmosphere by means of Joule heating of atmosphere by the electric currents proposed by Makarova et al. [14].

2. Quantitative Estimation of the Effects by the Electric Currents in the Middle Stratosphere

A new mechanism of the thermal heating in the middle stratosphere by the electric currents which connected with parameters of the global electric circuit will be demonstrated here. We are using the approach of Alfven and Falthammar [1] to evaluate the electric currents effects in a weakly ionized plasma.

Let us suggest that the density of the charged particles with charge e_k and mass m_k in atmosphere is equal to n_k. An electric field of strength E will move these particles with an averaged drift velocity u_k.

$$u_k = b_k E \tag{9}$$

where b_k is a constant, which can be called mobility in a case of weak fields. If the values of mobility of all atmospheric gases are known it is possible to calculate intensity of electric current I and of conductivity of atmosphere σ since $I = \sigma E$ give equations for the current (I) and conductivity (σ) in the following form [1]:

$$I = \sum n_k e_k u_k = \sum n_k e_k b_k E \tag{10}$$

$$\sigma = \sum n_k e_k b_k \tag{11}$$

An electrical resistance of the atmosphere R can be calculated as:

$$R = 1/\sigma = 1/\sum n_k e_k b_k \tag{12}$$

We evaluate the effects of electric currents in the atmosphere with a simplified method based on concept of “particle free path length”. The result of our calculations will depend on the parameters involved whose values are difficult to evaluate exactly.

Among these the frequency of ion-neutral collisions and the concentration of the charged particles are poorly known. “Particle free path length” approach assumes that the charged molecules undergo instant collisions while between the collisions they move with acceleration under the influence of electric field.

Alfven and Falthammar [1] defined the mean velocity u_k of a particle moving in the direction of electric field by the “particle free path length” method as:

$$u_k = (e_k \lambda_k / 2m_k V_k)\, E \text{ (m/s)}, \tag{13}$$

where $\lambda_k = V_k / \nu_k$ is a particle free path length; V_k is a mean thermal velocity of ions ; ν_k - is ion-neutral collision frequency.

Then the mobility of a particle follows from (1) as:

$$b_k = u_k / E = \gamma\, e_k \lambda_k / m_k\, V_k = \gamma\, e_k / m_k\, \nu_k \text{ (m/s)}, \quad (14)$$

where γ is a dimensionless coefficient.

Values of γ vary in the range between 0.5 and 1.0. An accurate calculation of the mobility taking into account statistical distribution of the velocities and the free path lengths gives the expression similar to equation (6) but with $\gamma = 1$ [1]. Inserting these values of mobility into the expressions for current intensity and resistance we obtain:

$$I = \Sigma(n_k e_k^2\, E)/\, m_k\, \nu_k \text{ (A/m)} \quad (15)$$

$$R = 1/\Sigma(n_k e_k^2\,)/\, m_k\, \nu_k = m_k/\Sigma(n_k e_k^2\, t_k) \text{ } (\Omega \text{ m}) \quad (16)$$

The charged component of the middle atmosphere consists of the electrons as well as positive and negative ions. As a rule, these particles are in the state of electric equilibrium when the concentrations of positive and negative particles are equal. In the equations (15) and (16) the summation should be made for negative (mainly electrons) and positive particles separately. The mass of the electron (m_e =1.6 10^{-31} kg) is much less than mass of the ion (m_i=M m_p=M 1.6 10^{-27}kg, where M is a mean mass number of ions). Their ratio m_e/m_i is equal to 1.57 10^{-6} for M=400 at stratospheric heights [5]. In this case the part of the equation for electron component of the current is a dominant in expression for current (15) while ionic component is a main part in expression for resistance (16).

The amount of Joule heating produced by current I is:

$$dQ/dt = I^2\, R \text{ (J /m}^3 \text{ s)}. \quad (17)$$

Inserting (7) and (8) into (9) yields:

$$dQ/dt= n\, e^2\, E^2\, m_i/\nu\, m_e^2 \text{ (J /m}^3 \text{ s)} \quad (18)$$

where $n=n_i = n_e$ – total concentration of the charged particles for electrodynamic equilibrium conditions.

The external parameters, which determine amount of the Joule heating, dissipated by the electric currents in the atmosphere (equation 10) are the following: electric field strength E, charged particle concentration n, collision frequency ν, and ion mass weight m_i.

3. Quantitative Calculation of Changes in Thermal Regime of the Stratosphere Caused by Electric Currents with Realistic Input Parameters

Accurate values of these parameters are difficult to evaluate. In this section we will try to collect as much information as possible from different sources.

Speaking about strength of the electric field in the stratosphere E there are very few reliable measurements of this parameter. Bering [4] performed one of them by using the special long-lived balloons in summer seasons of 1988-1989 over South Pole Station in Antarctica. The range of the measured electric field values was 0.1 - 0.35 V/m. We have sorted these data in accordance with the solar wind disturbances taken as the subsolar distance between the Earth and dayside magnetopause position expressed in the Earth radius units Re [12] and got the following empirical relation between these two parameters:

$$E = (527.01 - 31.5\ \mathrm{Re})\ 10^{-3}\ (\mathrm{V/m}). \tag{19}$$

This relation was derived for ~ 40 experimental values of stratospheric electric field. Calculated correlation coefficient between the magnetopause positions and stratospheric electric field values turned out to be rather high – 0.78.

This equation (19) is used for the calculation of electric field parameters. This choice is explained by the lack of other reliable measurements of stratospheric electric fields and must be considered as the first approximation of the global electric field distribution in the stratosphere. The magnetopause position is determined from an empirical model developed by Shue et al. [16]. The input parameters of this model are the solar wind density n_{sw} , velocity V_{sw} as well as the magnitude and orientation of the interplanetary magnetic field (IMF) vertical component B_z. These parameters are given in the NASA periodic publications such as the Solar-Geophysical Data. The magnitude of the solar wind impact on the Earth's atmosphere is maximal at the dayside of our planet during 06-18 hours of local time [11]. Physical processes responsible for interaction between the solar wind and the near-Earth space on the Earth night side (18-06 local time) have not been investigated properly yet. Therefore, the magnitude of the electric field strength induced by the solar wind on this part of the Earth could not be determined from the equation (11) and at the moment only very rough estimations could be made.

These estimations show that the magnitude of stratospheric electric field on the Earth's nightside is of about 40 percent of the corresponding daytime value.

As for concentration of the charged particles in the middle atmosphere (18-40 km) it is formed by many sources. The main source is the galactic cosmic rays (GCR), which permanently ionize this part of the atmosphere at all latitudes from the pole to the equator. Concentration of the free ions in the atmosphere is determined as proposed in [20]:

$$n_i = [q/\alpha_{eff}]^{1/2}, \quad (20)$$

where n_i is the ion density in m^{-3}; q is the rate of ionization in ion/m^3s; α_{eff} is the effective recombination coefficient in m^3s^{-1}. In general the latter parameter depends on altitude and the solar irradiance, but according to [20] we put it equal to 1.10^{-9} m^3 s^{-1}. Ionization rate (q) produced by the GCR fluxes could be calculated by the equation from [8]:

$$q= (A+B \sin^4\varphi)\, n_a, \quad (21)$$

where φ is the geomagnetic latitude; A and B are dimensionless coefficients depending on the solar activity level; n_a is the concentration of neutral molecules in the atmosphere. A is a constant coefficient equal to 1.74 10^{-18}, B is a coefficient equal to 1.93 10^{17} for the years of maximum solar activity and equal to 2.84 10^{-17} for the years of minimum solar activity. The solar cosmic rays, the solar X-rays and aerosols are able of producing an anomalous ionization at these altitudes, only sporadically though. The rate of ionization by the GCR depends on geomagnetic latitude (it increases from the equator toward the pole) and on the solar activity (it is highest/lowest during the solar activity minimum/maximum).

The vertical distribution of the charged particles in the atmosphere obtained from the experimental data has been presented in [19]. The appearance of the secondary ionization maximum at 24 km altitude presumably consisting of the heavy ion-clusters is clearly seen. Its density is about 3.0 10^9 m^{-3}. Another experimental data confirming the presence of similar layer in the stratosphere can be found in [5]. Thus, there are solid reasons to suggest the existence of a charged particles layer at stratospheric altitudes for any level of solar and geomagnetic activity. It is a very important point for the present investigation.

Frequency of the ion-neutral collisions (ν_k) is an important parameter of the middle atmosphere determining electrodynamics of this region. In general form it could be calculated by the following equation [20]:

$$\nu_k = 4.4\ 10^{-15}\,(T/300)\ n_a\ (s^{-1}) \tag{22}$$

where T is the atmospheric temperature in K; n_a is the density of the neutral molecules in the atmosphere in m^{-3} . There are few experimental measurements of this parameter at altitudes 18-40 km [20] which gave the value of ν_k equal to 1.5 10^9 s^{-1} at the 30 km altitude.

One of the main points of our investigation is the established fact of existence of a layer of heavy ions at altitudes ~25-35 km [5, 19]. This layer consists of the cluster ions with the mean mass weight of around 400. This value is taken as universal parameter in further studies.

4. Numerical Estimation of the Heating Rate in the Middle Atmosphere Produced by the Electric Currents

Using the parameterization of all components in the final equation (10) it is possible to make a numerical estimation of thermal effects produced by the electric currents in the atmosphere. For this calculation we have applied moderately disturbed conditions in the solar wind. For this particular case we can take the following values of the controlling parameters:

m_p	(proton mass)	=	1.6 10^{-27} kg;
m_e	(electron mass)	=	9.1 10^{-31} kg;
e (	charge of electron)	=	1.6 10^{-19} C
E		=	3 10^{-1} V/m
T		=	220.0 K
n_i		=	4 10^8 m^{-3}
M		=	400
ν_k		=	1.51 109 s-1

By substituting these values into the equation (10) we can obtain the amount of Joule heating equal to 1.8 .10-4 J/m3s. For the unit of volume (if we take c_p = 240 kal/kg K and ρ = 4. 10-2 kg/m-3) we got dT/dt = 3. 10-2 K/ hour.

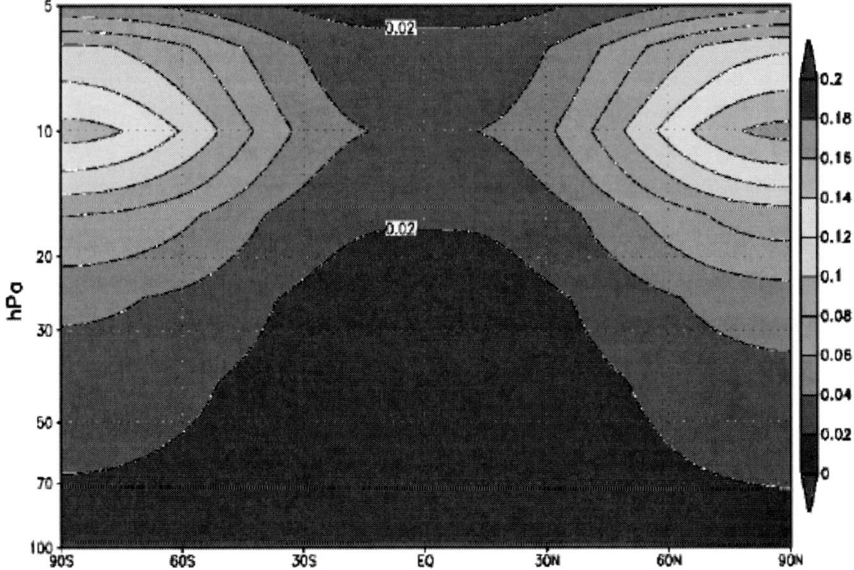

Figure 2. Zonal mean heating rate (K/day) averaging over all longitudes for every value of latitude due to solar wind induced electrical currents in the middle stratosphere. The heating rates have been calculated using the parameterization and input data for January 1996.

It is known that similar impact of the solar UV radiation on the ozone layer produces $dT/dt = 0.5\ 10^{-2}$ K/day. If the maximal value of the charged particles density in the stratosphere ($n_i = 3.0\ 10^9\ m^{-3}$) is used in this calculation we get $dQ/dt = 3.0\ 10^{-4}\ (J/m^3s)$. It means that the temperature tendency due to proposed mechanism would be equal to $dT/dt = 1.2$ K/day. The latter value is comparable with correspondent atmospheric heating due to solar UV radiation [7]. Results of the similar calculations for various situations showed the same order of Joule heating due to the solar wind dynamics as those described above. Figure 2 illustrates zonal mean distribution of the Joule heating rate. The calculations were performed for January and for moderate level of solar and interplanetary activity by using the method described above. The geographical distribution of the Joule heating rate at 10 hPa surface due to the solar wind dynamics is shown in Figure 2.

The heating rate increases monotonically from the equator to the poles in accordance with the ionization rate (see equation 21). The heating rate reaches its maximum over high latitudes of the summer hemisphere (Antarctica) since density of the neutral atmosphere is higher under such circumstances than in the winter. The both ionization rate and frequency of ion-neutral collisions crucially depend on this parameter (see equations 21 and 22). As one can see the magnitude of the Joule heating given in Figures 2 are about 3 times lower than the corresponding maximum values indicated earlier. This difference is explained by a moderate level of solar activity adopted for the calculations with correspondingly smaller ionization rate and density of the neutral atmosphere. The most important result of these calculations is that even these values of the Joule heating caused by the solar wind dynamics are comparable with correspondent atmospheric heating rates due to solar UV radiation at these heights [7].

The direct consequences of such warming could be the changes in dynamics of the stratospheric polar vortex, global ozone concentration and also climate/weather pattern. All these processes could only be accurately evaluated using a global scale 3D chemistry-climate model and the parameterization for the additional heating presented and explained here.

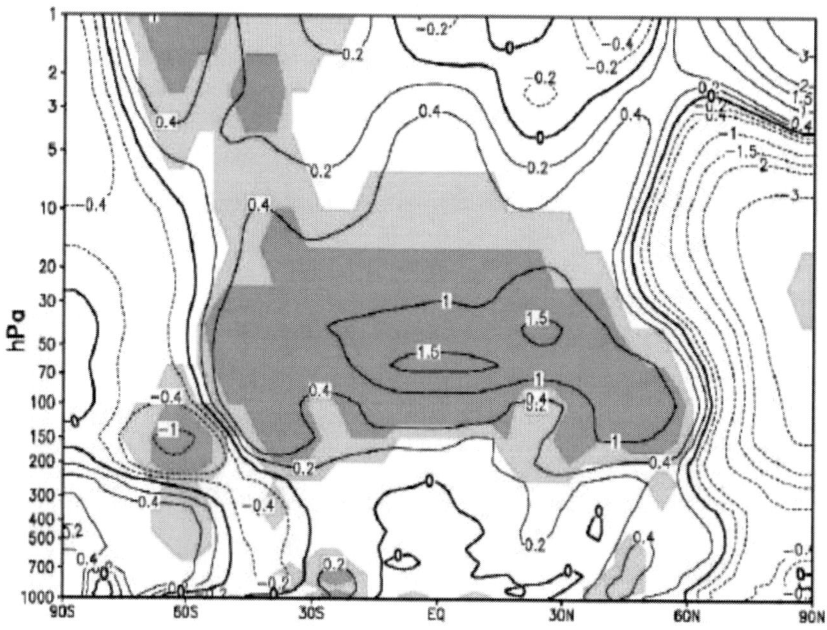

Figure 3. Simulated changes of the monthly zonal mean (a) temperature, in K in %, due to Joule heat forcing for March. The shading indicates the level of the statistical significance (light – at or more 80%; heavy – at or more 95.

A Chemistry Climate model [23] is used to evaluate of the possible influence of Joule heating induced by the solar wind and interplanetary magnetic field elements on the ozone concentration and dynamics of the Earth stratosphere. The Joule heating rates in the stratosphere are parameterized on the base of the time series of the solar wind and IMF parameters taken from NASA satellite dataset (1973-1999) for 1996.

5. Influence Conductivity of the Ground Earth as One Main Parameter of the Global Electric Circuit

The stratospheric warming as a result of the increasing energy of the solar wind and magnetopause approaching the Earth was explained by Makarova and Shirochkov [12] in the framework of a modified version of the global electric circuit with an external electromotive force (EMF) generator.

In this case some part of the solar wind energy is transferred into the system “magnetosphere – ionosphere – atmosphere” as the energy of induced electric fields.

The Earth’s stratosphere where a layer of increased heavy ion concentration is found [15] could be one of the conducting layers of the global electric circuit under certain circumstances. Furthermore, electric currents flowing through this stratospheric ion layer will produce Joule heating of the stratosphere. The reality of this hypothesis was proved by numerical calculations, which showed that this process could produce the observed warming of the stratosphere by several degrees [14]. The next step was the inclusion of this algorithm into one of the few existing global models of the atmosphere. Numerical experiments with this modified version of the model demonstrated efficiency of the proposed mechanism variability of stratospheric temperature and the ozone density [23].

Using a modified version of the global electric circuit mentioned earlier an increase of the solar wind dynamic pressure will produce more intense electric currents in the circuit and, consequently, a warming of the stratosphere. The observed phenomenon of cooling of the stratosphere could be explained by a re – distribution of the currents in the circuit due to change of electric conductivity of separate elements of the circuit. The electric conductivity of the stratosphere can generally be taken as $\sigma = 10^{-10}$ S/m. The conductivity of the ground surface is much greater - $\sigma = 2 - 10^{-2}$ S/m. The latter values are typical for seawater and cultivated soil. It is well known that the ground surface in the Arctic is typically either snow above ice or permafrost. These kinds of the ground surface are not good electric conductors since typical values of the conductivity of permafrost soil is $\sigma = 10^{-7}$ S/m while for the dry ice it is $\sigma = 10^{-8-}$S/m. A layer of increased ion concentration could exist in the stratosphere at altitudes 20 – 30 km. The ion density in this layer could be as high as $n = 5\ 10^{9}\ m^{-3}$ [15] which means that stratospheric conductivity here could be $\sigma = 10^{-7}$ S/m. By comparison, ice conductivity is equal to $\sigma = 10^{-8}$ S/m under air temperature of -10°C while it becomes equal to $\sigma = 5\ 10^{-10}$ S/m under air temperature minus 40°C [22]. Therefore the situations could be expected when conductivity of the ground surface is close to the values of σ observed in the stratosphere.

The data of atmospheric balloon sounding at four arctic and subarctic stations were chosen for study of the variations of atmospheric temperature as functions of the solar wind energy and results are presented in Table 1.

Data of Table 1 shows examples of the dependence of stratospheric temperature on the solar wind dynamic pressure at several stations. Both parameters are taken for each December, September and June in the period from 1964 to 2002. Data of Table 1 demonstrate that the increased input of solar wind energy could cause a warming (positive correlation for Lulea) as well as a cooling of the stratosphere (negative correlations for Barter, Ziryanka, Fairbanks). Rather unusual data were obtained for Murmansk: the temperature of the stratosphere at this station could be either directly proportional to an increase of the solar wind or inversely proportional to it. Due to this, the correlation coefficient between these two parameters appears to be very low at this station. One can see that the correlation between stratospheric temperature and solar wind dynamic pressure for Barter is negative for all seasons, while it is always positive for Lulea. The same parameter for Fairbanks and Ziryanka changes its sign with the seasons (it is negative in winter and positive in summer). However, the data of Murmansk demonstrate its two-fold character of the relation between these parameters for all stations of year. Therefore the resulting correlation coefficient between solar wind dynamic pressure is very low for Murmansk in every season of the year.

It seems that the results obtained could be explained in the framework of the global electrical circuit concept under the assumption that changes of the electric conductivity of the stratosphere and the ground surface must redistribute the electric currents in the elements of the global electric circuit. In this case the seasonal variations of the obtained relations (see Table 1) could be attributed to natural seasonal changes of the conductivity of the ground surface, e.g., how wet it is, or how covered with ice.

The conductivity of the Arctic stratosphere could be higher than of the ground surface under the same circumstances. In this case the electric current in the global electric circuit will flow through the stratosphere under undisturbed conditions in the solar wind since the corresponding conductivity of the ground surface (ice) will be lower than in the stratosphere. As a result of such a redistribution of the currents the stratosphere will be warmed. This phenomenon is illustrated well by the data of Table 1 where it is shown that under quiet conditions in the solar wind (low dynamic pressure) the stratospheric temperature is higher at the high – latitude stations: Barter and Fairbanks than at the sub – auroral station Lulea.

One must take into account that the conductivity of the ground surface covered by ice could increase with increasing air humidity and temperature. In this case a re–distribution of the currents inside the global electric circuit takes place. Under increased solar wind dynamic pressure the currents flowing through the stratosphere will be less than the currents flowing through the ground surface.

Consequently cooling of the stratosphere with an increase of the solar wind dynamic pressure can be explained. Confirmation of this effect can be seen in the data of Table 1 where a negative correlation between these two parameters is shown. The change of the sign of the correlation between the solar wind dynamic pressure and stratospheric temperature is connected closely with the seasonal variations of the ground surface conductivity.

Zakharov [22] demonstrated the maps of the ice distribution in the Arctic for different seasons. The areas covered by permanent ice, as well as by drifting ice were shown on these maps. These maps showed, that ground surface at Barter station (northern coast of Alaska) is covered by permanent ice for the whole year. Quite naturally we have the negative correlation between the solar wind dynamic pressure and stratospheric temperature (see Table 1).

At the stations Fairbanks and Zirynka located at lower latitudes the ground surface thaws in summer, therefore the negative correlation sign for the winter months is transformed to positive values in summer.

The sub auroral station Lulea is located beyond the permafrost zone; thus the ground surface conductivity here does not change significantly with season. The stratospheric temperature at this station will depend primarily on the level of the solar wind disturbance.

Table 1. Correlation coefficients between stratospheric temperature at 50 hPa isobaric surfaces and the solar wind dynamic pressure for 1964 –2002 years

Station	Geographic latitude	Geographic longitude	December	September	June
Barter	70. 13° N	216. 36° E	-0.71	-0.7	0.65
Fairbanks	64. 81° N	212. 13° E	-0.8	0.54	0.49
Ziryanka	65. 73° N	150. 89° E	-0.72	-0.54	0.51
Murmansk	68. 96° N	33. 04° E	-	-	-
Lulea	65.54° N	22.13° E	0.75	0.71	0.54

The greater the solar wind dynamic pressure the warmer will be the stratosphere at this station. The different situation in ground surface conductivity is observed at Murmansk in all seasons where either approaching sea ice or open seawater are presented. Therefore we obtained the complicated dependence between the stratosphere temperature and the solar wind dynamic pressure at this station. So, variations of temperature of the middle atmosphere are connected with both EUV radiation and solar wind energy. Observed results can be explained in framework of global electrical circuit driving by thunderstorm activity and energy of the solar wind. Temperature of stratosphere is increasing during disturbances of solar wind if conductivity of the Earth's surface is high. And vice versa-temperature of stratosphere has negative correlation with energy of the solar wind above places covered by ice. Middle atmosphere is warmer above places covered by ice than the surface with good conductivity during quite solar wind. Situation is opposite during disturbances of solar wind.

The Solar Wind Energy as a Part of the Total Solar Activity

Other data similar to those shown on Figure 1 demonstrate that the short-term variations of the stratospheric temperature are under influence of the solar wind energy changes [10, 12]. Furthermore it was found that the same relation could be traced in the long-term (several cycles of the solar activity) variations of stratospheric temperature in the polar atmosphere [13]. The latter authors showed that enhanced solar wind dynamic pressure caused warming of the stratosphere above both the Northern and Southern geographic poles while quite opposite effect was observed above Antarctic station Vostok located at distance about 1500 km from the South Pole. All these relations are found for winter months i.e. for the "polar night" conditions when the solar UV radiation impact on the polar atmosphere is minimal.

Since the solar wind energy is proved to be geoeffective force, it is worth trying to determine its relation to other kinds of the solar activity whose influence on the Earth weather and climate is commonly accepted.

Figure 4 shows time series of the annually averaged values of the Sun spot numbers and the annually averaged values of the solar wind dynamic pressure for period of 1970-2008.

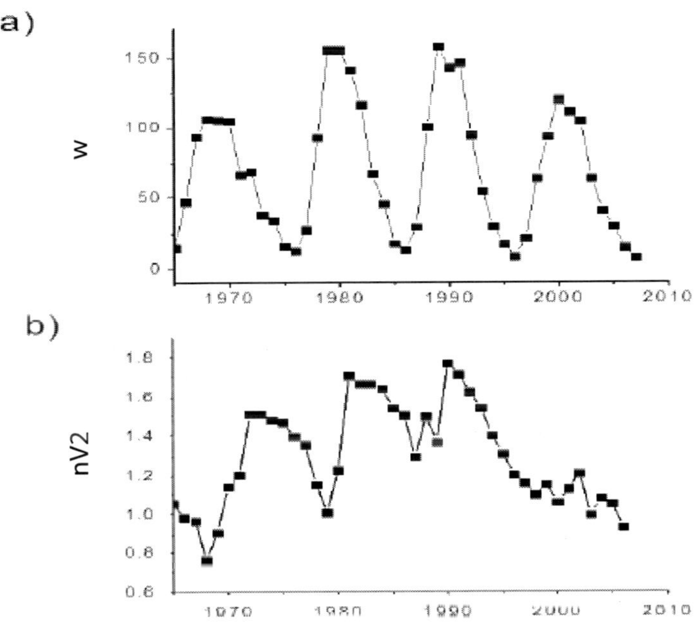

Figure 4. Time of various parameters of the solar activity (annually averaged values f W) (a) and the solar wind dynamic pressure (annually averaged values of nV2, nPa) (b).

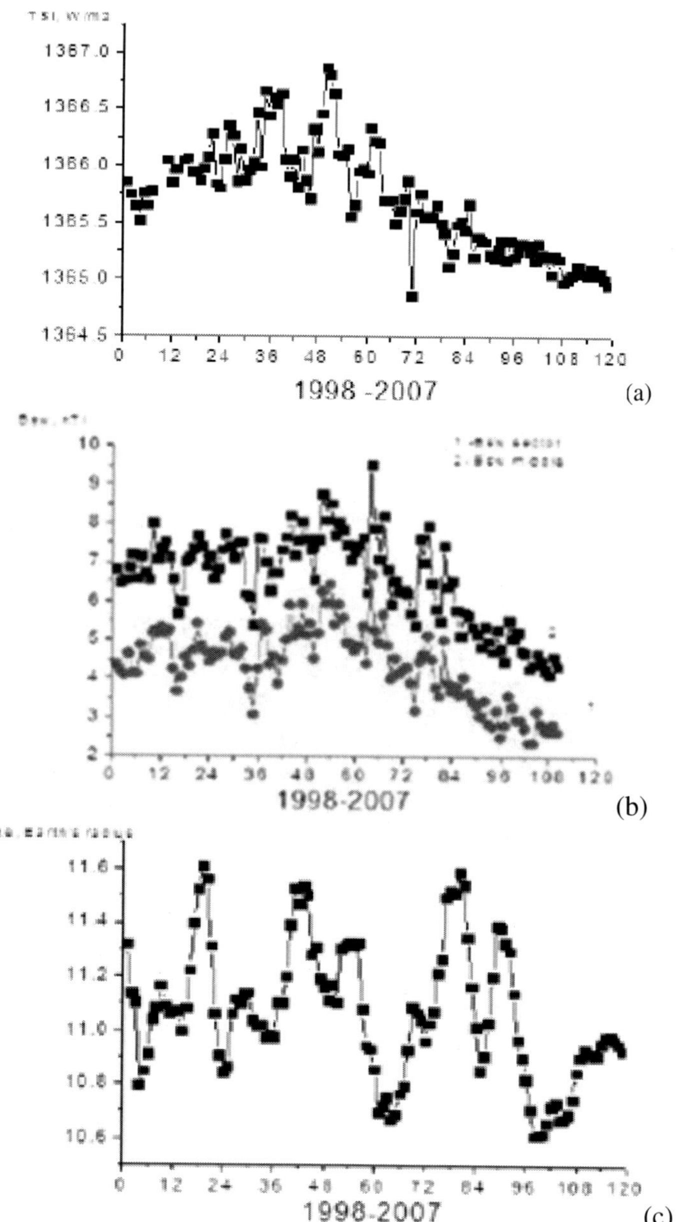

Figure 5. The long term variations a) the monthly averaged values of the total solar irradiance (TSI, Wm –2); b) the monthly averaged values of the magnetic field of the solar wind (B, nTl) and c) the monthly averaged values of the position magnetopause.

One can see that the both data sets vary synchronously during period of 1970-2000 when their extremes coincide in time. However, this similarity disrupted suddenly in the last 23d cycle of the solar activity. If the sunspot number W continue to vary in accordance with usual 11-year period the solar wind dynamic pressure demonstrate monotonous diminishing. In order to explain this unusual solar anomaly we considered temporal variations of other important solar parameters shown at Figure 5.

The quasi-stationary variations of the parameters of the solar wind during 23 cycle of solar activity were studied. It was found that the main peculiarity of this cycle is diminishing magnetic field of the solar wind as compared with the previous two cycles. Maximum of magnetic field of the solar wind as well as many high velocity fluxes with low density of plasma is observed in 2003 -2004 after phase of sunspot maximum in June of 2002. The high velocity solar fluxes with low density plasma are connected with increasing of corona mass injections during early declining phase of the solar *activity*. The main peculiarity of the 23 cycle of the Sun activity is diminishing magnetic field of the solar wind and TSI on the end of cycle, and quasi-stationary changes position of the magnetopause position as indicator of electromagnetic energy of the solar wind. The quasi-stationary variations of the parameters of the solar wind during 23 cycle of solar activity are shown on Figure 5. The quasi-stationary values of magnetic field of the solar wind during minimum solar activity in 2005 -2007 years became to be equal to 4 nTl while it was approximately 6 nTl in the previous 21 and 22 cycles. Non - stationary and dense fluxes of the solar wind with values of velocity approximately 450 km/s are characteristic for minimum solar activity in the 23 cycle. It is reasonable to suggest existence of connection between these fluxes with development of heliospheric current sheaths and coronal streamers when large magnetic fields of the Sun are not strong.

The quasi-stationary values of magnetic field of the solar wind shows some more long cycle of variation which is approximately 50 years with maximum in 21 and 22 cycles and minimum in 20 and 23 cycles. Structure of magnetic fields of the Sun determines different kinds of the solar activity including intensity of total solar irradiance (TSI) and energy of the solar wind transferred into the near- Earth space.

One of the main products of energy of the solar wind is the electric field system, which is observed in the ionosphere, middle atmosphere and on the ground surface. Therefore it could influence climate of the Earth. Knowledge of the real periodicity of the solar activity is very important for human activity.

Conclusion

1. Temporal variations of the temperature of the middle atmosphere are connected with the solar wind energy.
2. The results observed can be explained in the framework of the global electrical circuit driven by energy of the solar wind.
3. The temperature of the stratosphere increases during disturbances of the solar wind if the conductivity of the Earth's surface is high. Vice versa-the temperature of the stratosphere above places covered by ice has a negative correlation with energy of the solar wind.
4. The stratospheric layer with increased concentration of ions created by galactic or solar charged particles is an important element of the global electric circuit in the polar regions.
5. Long –term variations of the different kind of the solar activity can influence on electromagnetic energy and fluxes of galactic or solar charged particles which enter the near - Earth Space and change the Earth climate and weather.

Acknowledgment

This work was performed partly due to assistance of the Russian Basic Research Foundation (RBRF) Grant 06-05-64311.

References

[1] Alfven, H., Falthammar, C. G. *Cosmical electrodynamics*; Clarendon Press, Oxford, U. K., 1963, 5-280.

[2] Baker, D. N. J. Atmos. *Solar-Terr. Physics*. 2000, vol. 62, 1669 – 1681.

[3] Bazilevskaja, G. A., Krainev, M. B., Makhmutov, V. S. J. Atmos. *Solar-Terr.Physisc*, vol.62, 2000, 1577-1588.

[4] Bering, III, E. A., Few, A. A., Benbrook, J. R. *Physics Today*. 1998, vol. 51(10), 24-30.

[5] Brasseur, G., Chatel, A. *Ann. Geophys*. 1983, vol. 1, 173-185.

[6] Callis, L. B., Nataranjan, M., Lambeth, J. B., Baker, D. N. *J. Geophys. Res*. 1998, vol. 103, 28421- 28438.

[7] Ginzburg, E. I. In: *Construction principles of the upper atmosphere dynamics models;* Editor Ginzburg, E. I., Hydrometeoizdat, Moscow, U. S. S. R., 1989, 3-121 (in Russian).

[8] Heaps, M. G., Planet. *Space Sci.* 1978, vol. 26, 513-517.

[9] King, J. H., Couzens, D. A.. *Interplanetary Medium Data Book-Supplement 3A*, N. A. S. A., Washington D. C., US, 1986.

[10] Makarova, L. N., Shirochkov, A. V., Grigor'eva, J. A., Volobuev, D. M., *Geomagnet. and Aeronomy*, 1997, vol. 37, 158- 164 (in Russian).

[11] Makarova, L. N., Shirochkov, A. V., Koptjaeva, K. V., *Geomagnet. and Aeronomy*, 1998, vol. 38, 159-162 (in Russian).

[12] Makarova, L. N., Shirochkov, A. V. 2000, *Phys. Chem..Earth*, 2000, vol. C 25, 495-498.

[13] Makarova, L. N., Shirochkov, A. V. *Phys. Chem. Earth,* 2002, vol. C 27, 449-453.

[14] Makarova, L. N., Shirochkov, A. V., Nagurny, A. P., Rozanov, E. V., Schmutz, W. *J. Atmos. Solar-Terr. Physics*, 2004, vol. 66, 1173-1177.

[15] Rosen, J. M., Hofmann, D. J. *J. Geophys. Res.*, 1988, vol. 93A, 8415-8422.

[16] Shue, J. H., Chao, J. K., Fu H. C. et al., *J. Geophys.* Res. 1997, vol. 102A, 9497-9506.

[17] Solanki, S. K., Fligge, M. *Geophys.Res.Lett.* 1998, vol. 25, 341-344.

[18] Troshichev, O. A., Egorova, L. V., Vovk, V. Ya. *Adv. Space. Res.*, 2004, vol. 34, 1824-1829.

[19] Volland, H. In: *Modern Ionospheric Science;* Editors Kohl, H., Ruster, R., Schlegel, K., Max-Planck-Institut fur Aeronomie, 37191 Katlenburg-Lindau, F. R. G., 1996, 102-135.

[20] Webber, W. *J. Geophys. Res.* 1962, vol. 67, 5091-5106.

[21] Whitten, R. C., Poppoff, I. G. *Physics of the lower ionosphere*; Prentice-Hall, Inc., Englewood Cliffs, N. J., US, 1965, 3-292.

[22] Zakharov, V. F. *Sea ice in climatic system.* Hydrometeoizdat, Saint-Petersburg, Russia, 1996, 3-213 (in Russian).

[23] Zubov, V. A., Rozanov, E. V., Shirochkov, A. V., Makarova, L. N., Egorova, T. A., Kiselev, A. E., Ozolin, Y. A., Karol, I. L., Schmutz, W. *J. Atmos. Solar-Terr. Physics*, 2005, vol. 67, 155-162.

INDEX

A

acceleration, 118
activity level, 109, 121
aerosols, 104, 121
Africa, 59
air, 126, 127
Air Force, 103
air temperature, 14, 30, 32, 64, 126
Alaska, 127
algorithm, 125
amplitude, 64, 83
anomalous, 121
Antarctic, 107, 129
Arctic, ix, 107, 110, 125, 127
Asia, 63
asymmetry, 3, 41
atmospheric disturbances, vii, 2
atmospheric pressure, viii, ix, 14, 41, 45, 47, 50, 51, 73, 88, 89, 92, 102
atoms, 12
attribution, 54
averaging, 123

B

base, 125
behavior, 110
biosphere, 9
Boltzmann constant, 116
Brazil, 67, 70
breakdown, 29
Britain, 38
Bulgaria, 44

C

Cairo, 72
Canada, 114
carbon, 66
case study(ies), 32, 41
casting, 9
causality, 18, 30
centre of gravity, viii, 47, 49
charged particle, 114, 115, 117, 118, 119, 121, 123, 132
chemical, x, 16, 108, 114
chemical characteristics, 16
Chicago, 68, 71
circulation, viii, 3, 18, 19, 29, 35, 38, 40, 41, 47, 48, 50, 52, 53, 54, 55, 56, 57, 59, 60, 62, 64, 65, 66, 73, 80, 81, 101, 102
City, 44
classes, 13
climate, vii, viii, 10, 14, 30, 40, 45, 47, 50, 51, 54, 55, 59, 60, 61, 62, 63, 64, 65, 66, 74, 103, 104, 109, 112, 124, 129, 131, 132
climate change, 10, 30, 40, 62, 63, 74
clusters, 113, 114, 116, 121

CO_2, 64
coherence, 14
collisions, 15, 116, 117, 118, 122, 124
combined effect, 87
commercial, 38, 63
community, 68
components, 115, 117, 122
composites, 66
composition, 113
concentration, 114, 115, 118, 119, 121, 124, 125, 126, 132
conception, vii
conductivity, x, 100, 101, 108, 113, 118, 125, 127, 132
conductor(s), 19, 126
conference, 61
confidence, 112
confidence interval, 112
configuration, ix, 52, 108
Continental, 103
contour, 18, 19
control, 109
cooling, 55, 77, 78, 80, 82, 89, 102, 125, 126, 127
corona, 131
correlation, viii, 3, 14, 41, 47, 49, 50, 53, 54, 56, 74, 83, 86, 94, 112, 113, 120, 126, 127, 128, 132
correlation coefficient, 14, 120, 126
correlations, 3, 53, 55, 57, 60, 126
cosmic ray flux, 3, 45, 75, 104, 115
cosmic rays, 41, 44, 45, 50, 55, 56, 60, 69, 71, 74, 75, 102, 103, 104, 114, 115, 121
Coulomb, 116
coupling, ix, 108, 110, 113
covering, 5
cycles, 3, 35, 48, 49, 55, 56, 62, 66, 112, 128, 131
cyclogenesis, viii, 2, 14, 16, 27
cyclones, vii, 1, 3, 4, 14, 16, 18, 27, 28, 35, 37, 38
cyclonic motions, vii, 1, 2, 4, 6, 16, 38, 39

D

data set, 112, 131
database, 4
decay, 80, 94
density, 114, 115, 118, 120, 121, 122, 123, 125, 126, 131
descending air masses, viii, 73, 91, 92, 102
deviation, 77, 78, 79, 80, 83, 86, 89, 91, 98
dielectric constant, 116
direct measure, 70
dislocation, 51
distribution, 5, 6, 9, 13, 17, 45, 50, 51, 58, 92, 97, 112, 114, 115, 119, 120, 121, 123, 125, 127
divergence, 3
drainage, 80, 86, 92, 103
dry ice, 126
duration, 109

E

Earth magnetosphere, ix, 108
earthquakes, 53
Egypt, 72
El Niño, 56, 64, 94, 97
electric circuit, ix, 108, 110, 112, 113, 117, 125, 126, 127, 132
electric conductivity, x, 108, 125, 126
electric current, 99, 100, 112, 113, 117, 118, 119, 122, 125, 126, 127
electric energy, 113
electric field, viii, 3, 42, 73, 75, 86, 89, 99, 101, 102, 112, 113, 117, 118, 119, 120, 125, 131
electrical resistance, 118
electricity, 45, 104
electromagnetic, ix, 3, 16, 19, 21, 44, 107, 109, 131, 132
electromotive force, 125
Electro-Motive Force (EMF), ix, 108
electron, 114, 116, 117, 119, 122
electrons, vii, 11, 12, 16, 19, 20, 31, 32, 71, 116, 117, 119

EMF, ix, 108, 110, 125
emission, vii, viii, 16, 47, 48, 49, 53, 54, 55, 59, 68
energy, vii, viii, ix, 2, 3, 4, 5, 6, 7, 8, 11, 12, 15, 16, 18, 19, 27, 30, 36, 38, 39, 44, 49, 55, 64, 71, 74, 107, 109, 110, 112, 113, 116, 117, 125, 126, 128, 129, 131, 132
energy transfer, 109
England, 28, 35
environment(s), viii, 9, 16, 30, 35, 37, 41, 47, 50, 51, 54, 60, 65
equilibrium, 27, 82, 101, 115, 119
Europe, 22, 24, 28, 29, 44, 55, 58, 59
evidence, 2, 15, 44, 45, 62, 66, 77, 104, 110
exposure, 86

F

Fairbanks, 126, 127, 128
fax, 107
February, 114
fires, 41, 44
flow, 127
fluctuations, ix, 34, 35, 41, 61, 65, 74, 94, 102
fluid, 3
force, 18, 19, 20, 21, 91, 92, 110, 112, 125, 129
forecasting, 9, 43
forest fire, 41
formation, viii, 3, 16, 35, 36, 38, 50, 54, 55, 59, 70, 73, 77, 79, 80, 82, 92, 97, 101
formula, 91
freezing, 101
friction, 19, 52

G

galactic, 121, 132
gas, 116
gases, 118
generators, ix, 108, 113
geology, 66
geomagnetic field, 115
geomagnetic substorms, ix, 107, 109
geometry, 71
Georgia, 29
Germany, 68
global climate change, 62
global scale, 124
global warming, 3
GOES, 30
GPS, 45
graph, 58, 59
gravitational effect, 37
gravitational stress, 62
gravity, vii, viii, 11, 36, 47, 49, 50, 52, 60
growth, 34, 51, 80

H

Hawaii, 61
heat, 124
heating, x, 108, 112, 117, 119, 122, 123, 124, 125
heating rate, 123, 125
height, 66, 74, 87, 88, 90, 112, 114, 115
hemisphere, 3, 4, 5, 20, 24, 95, 103, 104, 105, 114, 123
Highlands, 30
history, 69, 71
Holocene, 61, 63, 65
human, 131
human activity, 131
humidity, 7, 127
Hurricane Katrina, 42
hurricanes, vii, 1, 2, 4, 5, 6, 7, 8, 9, 10, 13, 14, 23, 30, 35, 36, 39, 42
hydrogen, 11
hydrosphere, 6, 56, 57, 60
hypothesis, 16, 30, 35, 38, 74, 125

I

ice, 125, 127, 128, 132, 133
image(s), 26, 28, 30
IMF, 9, 15, 18, 43, 75, 76, 77, 78, 79, 80, 86, 92, 93, 94, 97, 101, 105, 120, 125

impulses, 99
incidence, 16, 29
inclusion, 125
income, 101
India, 1
indication, 110, 114
indices, 109
individuals, 69
induction, 18, 19, 21, 29
injections, 131
interaction, x, 108, 114, 115, 117, 120
interplanetary, 120, 123, 125
interval, 109
ion-clusters, 113, 121
ionic, 119
ionization, 11, 39, 75, 101, 115, 121, 123
ionosphere, 109, 113, 125, 131, 133
ions, ix, 108, 114, 116, 117, 118, 119, 121, 122, 132
Ireland, 22, 23, 24, 25, 26, 27, 45
iron, 12
irradiation, 8, 101
Italy, 41

J

Joule heating, x, 108, 112, 117, 119, 122, 123, 124, 125
Jupiter, 49
justification, 30

K

kinetic energy, 116
King, 133

L

law(s), 27, 109
lead, viii, 47, 50, 51, 53, 54, 59, 74, 92, 102
life cycle, 38
lifetime, 27
light, 49, 71, 124
long period, 114
Louisiana, 29
low temperatures, 30
luminosity, viii, 47, 48, 49, 53, 54, 55, 60
lying, 11

M

magnetic, 120, 125, 130, 131
magnetic field, 4, 9, 10, 11, 12, 13, 15, 16, 18, 19, 21, 30, 38, 43, 44, 68, 71, 75, 76, 77, 78, 80, 83, 102, 120, 125, 130, 131
magnetosphere, viii, ix, 10, 13, 15, 37, 39, 43, 47, 49, 50, 51, 53, 54, 55, 56, 59, 60, 62, 68, 69, 75, 93, 99, 100, 101, 108, 109, 125
magnitude, 77, 116, 120, 124
majority, 27, 38
man, 77, 78
mantle, 51, 53, 55, 64
marine environment, 57
mass, 4, 5, 11, 15, 18, 21, 23, 26, 28, 42, 48, 49, 59, 64, 80, 81, 86, 113, 118, 119, 122, 131
measurement(s), 77, 78, 83, 89, 111, 114, 120, 122
media, vii, 2
memory, 69
meridian, 8, 34
methodology, 38
Mexico, 44
Ministry of Education, 39
mixing, 18
mobility, 118, 119
modeling, 110
modelling, 9, 12, 15, 42
models, x, 38, 66, 74, 108, 125, 133
molecules, 12, 115, 116, 118, 121, 122
momentum, 49, 55, 56, 57, 60, 61, 64, 65
Moon, 42, 49, 51, 52, 64
Moscow, 133
movement, x, 108, 116

N

NASA, 120, 125, 133
natural, 127
neutral, 11, 115, 116, 117, 118, 121, 122, 123
Ni, 114, 115, 117
normal distribution, 31
North America, 27
nucleation, 15
nuclei, 11, 12
nucleons, 16
nuclides, 54, 61

O

observations, 109
oceans, 5, 36
orbit, 15, 68, 75
organize, 27, 29
orientation, 120
oscillation, 3, 49, 66, 98, 99, 103, 104, 105
ozone, x, 108, 113, 114, 123, 124, 125

P

Pacific, ix, 4, 5, 9, 14, 30, 35, 36, 37, 40, 41, 56, 57, 65, 74, 93, 94, 102, 105
parameter, 109, 120, 121, 122, 124, 126
particles, 114, 115, 116, 117, 118, 119, 121, 123, 132
passive, ix, 108, 113
periodic, 120
periodicity, 14, 57, 63, 131
permafrost, 125, 127
Philippines, 30
phone, 107
photochemical, 117
physical phenomena, 9
physics, 3, 4, 45, 109
planets, 48, 49
plasma, 117, 131
polar, ix, 11, 13, 16, 19, 61, 74, 77, 99, 100, 101, 102, 104, 107, 110, 124, 128, 132
polarity, 12
pollution, 7
porosity, 15
Portugal, 1, 42, 59
positive correlation, 126
precipitation, 3, 36, 112
pressure, ix, 107, 109, 112, 125, 126, 127, 128, 129
principles, 133
probability, 10, 27, 30, 36
production, 114
project, 61
propagation, 37, 38, 95
protons, vii, 11, 12, 16, 20, 24, 25, 27, 28, 31, 32, 34

Q

quartile, 10
quasi-equilibrium, 82, 101

R

radiation, viii, x, 3, 4, 5, 6, 7, 9, 10, 12, 15, 34, 68, 69, 71, 73, 77, 78, 79, 80, 82, 102, 108, 109, 123, 124, 128, 129
Radiation, 40, 43
radio, 71
radius, x, 18, 19, 21, 22, 91, 108, 110, 111, 112, 115, 116, 120
rainfall, 35
range, 119, 120
readership, 68
reality, ix, 69, 75, 108, 110, 113, 117, 125
recall, 6, 9, 14
recognition, 68
recombination, 115, 121
reconstruction, 54
redistribution, viii, 47, 51, 127
regeneration, 27
regression, 86
regression line, 86
regular, 109
resistance, 118, 119

resolution, 59
response, 15, 45, 60, 65, 75, 77, 86, 88, 89, 92, 104
Russia, 73, 107, 133
Russian, ix, 107, 132, 133

S

satellite, 109, 125
Saturn, 49
scale system, viii, 73, 80, 101
scatter, 75
scattering, 7
science, 68
scientific publications, 3
sea ice, 128
sea level, viii, ix, 14, 47, 51, 56, 57, 62, 63, 64, 65, 74, 80, 94, 102, 105
search, 109
seasonal changes, 127
seasonal variations, 126, 127
seawater, 125, 128
seeding, 36
sensing, 45
Serbia, 1, 39, 41
series, 125, 129
shape, 13, 69, 114
shock, 13, 42, 43
shock waves, 42
shoot, 23
short-term, 109, 128
showing, 35, 74
Siberia, 30
sign, 126, 127
signals, 53, 60, 65
significance level, 35
signs, 32
similarity, 131
simulation, 11, 62, 103
simulations, x, 11, 39, 108
SOI, 94, 95, 97, 98, 99
soil, 125
solar, ix, 107, 109, 110, 112, 113, 117, 120, 121, 122, 123, 124, 125, 126, 127, 128, 129, 130, 131, 132
solar activity, viii, 3, 9, 10, 14, 30, 31, 32, 34, 40, 41, 45, 47, 49, 50, 53, 54, 57, 60, 62, 74, 79, 109, 112, 121, 124, 129, 131, 132
solar energy, 109
solar influence, viii, 4, 47, 60, 61
solar system, 12, 49, 71
solar wind, ix, 107, 109, 110, 112, 113, 117, 120, 122, 123, 124, 125, 126, 127, 128, 129, 130, 131, 132
solar wind (SW), vii, 2
solution, 51
space environment, 9, 11, 50
species, 12, 113
speed of light, 11
St. Petersburg, 73
standard deviation, 31, 32
state(s), 9, 27, 29, 69, 82, 101, 119
storms, vii, ix, 1, 3, 9, 15, 35, 36, 38, 40, 68, 69, 107, 109
stratosphere, x, 108, 110, 112, 113, 114, 115, 116, 117, 120, 121, 123, 125, 126, 127, 128, 129, 132
strength, 118, 119, 120
stress, 30, 62
stroke, 30
structure, 10, 62, 82
substitution, 80
summer, 120, 123, 126, 127
Sun, vii, viii, ix, 2, 3, 4, 5, 7, 8, 9, 10, 11, 12, 13, 19, 23, 29, 30, 38, 39, 43, 47, 48, 49, 61, 62, 66, 67, 69, 70, 71, 99, 103, 107, 109, 129, 131
sunspot(s), 10, 11, 65, 109, 131
sunspot cycle, 10, 48, 49, 53
surface layer, viii, 73, 80, 102
Sweden, 47, 61

T

TCR, 29, 42, 44
technologies, vii
temperature, vii, ix, 10, 11, 14, 16, 25, 40, 41, 43, 54, 57, 62, 66, 69, 70, 74, 79, 82, 83, 84, 85, 86, 87, 88, 103, 104, 105,

107, 110, 111, 112, 116, 122, 123, 124, 125, 126, 127, 128, 132
temporal, 131
temporal variation, 131
terrestrial climate, vii, viii, 47, 50, 60, 62, 65
thermodynamics, 3
tides, 36
time, 113, 117, 120, 125, 129
time series, 17, 59, 125, 129
tornadoes, 24, 29, 30, 38, 39, 41, 44, 45
total energy, 74, 109
trajectory, 21, 24, 116
transport, 56, 59
transportation, 2
triggers, 11
tropical storms, 35
turbulence, 6

U

UK, 45, 132
uniform, 7
US, 45, 66, 68, 114
USSR, 133
UV radiation, x, 8, 42, 108, 109, 123, 124, 129

V

vacuum, 68
values, 115, 118, 119, 120, 122, 124, 125, 127, 129, 130, 131
variability, 109, 125
variables, 7, 31, 32, 35, 49
variation(s), 3, 7, 8, 35, 38, 41, 45, 48, 49, 50, 54, 60, 61, 63, 64, 69, 74, 75, 77, 86, 89, 102, 104, 112, 126, 127, 128, 130, 131, 132
vector, 18, 19, 91
velocity, 12, 17, 21, 22, 50, 69, 75, 116, 118, 120, 131
victims, 2
vortex, 124

W

war, 68
Washington, 66, 103, 133
water, viii, 5, 36, 45, 47, 56, 57, 58, 59, 94, 101, 104
weakness, 38
Western Europe, 26, 28
wind, ix, 107, 109, 110, 112, 113, 117, 120, 122, 123, 124, 125, 126, 127, 128, 129, 130, 131, 132
wind speed, 42, 92, 94
winter, 112, 124, 126, 127, 129
witnesses, 24
worldwide, 35
Wyoming, 114

X

X-rays, 121